AF885570

SOTHIS PRESS

Sothis Press aims to completely reimagine the public domain book with an emphasis on universal access to captivatingly designed print editions. We believe that technology will enable every single book ever published to be perpetually accessible in beautiful print form. Sothis is beauty in ubiquity.

www.sothis.press
1 Waterhouse Square, London, ECIN 2ST

An Elementary Treatise on Differential Equations

Abraham Cohen

SOTHIS PRESS

ISBN 978-1-0365-2399-2

An Elementary Treatise on Differential Equations
is in the public domain.

This edition is published by
Sothis Press © 2023. The company
is registered in England and Wales.
The Company Number is 14900389.

AN ELEMENTARY TREATISE

ON

DIFFERENTIAL EQUATIONS

BY

ABRAHAM COHEN, PH.D.

ASSOCIATE PROFESSOR OF MATHEMATICS
JOHNS HOPKINS UNIVERSITY

D. C. HEATH & CO., PUBLISHERS

BOSTON NEW YORK CHICAGO

PREFACE

THE following pages are the result of a course in Differential Equations which the author has given for some years to classes comprising students intending to pursue the study of Engineering or some other Physical Science, as well as those expecting to continue the study of Pure Mathematics or Mathematical Physics.

The primary object of this book is to make the student familiar with the principles and devices that will enable him to integrate most of the equations he is apt to come across. As much of the theory is given as is likely to be comprehensible to the student who has had a year's course in the Differential and Integral Calculus, and yet is sufficient to form a harmonizing setting for the numerous and otherwise apparently miscellaneous classes of equations, and the disconnected methods for solving them. It is intended to have the work sufficiently broad to make it a handy book of reference, without affecting its utility as a text-book. A number of footnotes and remarks have been put in, which, without breaking the continuity of the practical side of the subject, must prove of interest and value. Numerous historical and bibliographical references are also made.

A course that is limited in point of time and aims only at acquiring skill in integrating most of the equations that are apt to arise could dispense with §§ 12, 15, 17, 22, 28 (part in small type), 33, 34, 38–40, 46–48, 66–69 (except examples), 70, 71, 73, 75, 78, 80, 81. Many of these sections should properly come in a well-balanced course. The needs of the class, and the time at its disposal, must decide which of them, if any, should be omitted.

The author has had in mind continually the necessity of systematizing the various classes of equations that can be solved by elementary means, and of minimizing the number of methods by which they can be solved. To enable the student to get a better general view of the subject, the summaries at the ends of the various chapters and the final general summary must prove of great value.

Numerous applications to problems in Geometry and the Physical Sciences have been introduced, both in the body of the text and in the form of exercises for the student.

Although a large number of the problems have been published before, many are new, and all have been chosen to bring out the various methods of the differential equations, and of the integral calculus as well. Many of the examples worked out in the text were chosen to recall some of the more important methods of the latter; for while the use of tables of integrals is recommended, the student should not feel absolutely dependent upon them. Most of the solutions have a simple form or an interesting interpretation. Great care has been taken to avoid typographical errors. The author shall be very glad to learn of any that still exist.

The method of undetermined coefficients for finding the particular integral in the case of linear equations with constant coefficients is believed to be presented here for the first time in its complete form.

The subject of Partial Differential Equations is so vast that it was decided to present only a few topics, which, in all probability, will suffice for the needs of the students for whom this book is intended.

It was only after considerable thought that the author refrained from adding a chapter on the Lie Theory. It is hoped to present that important branch of the subject in a separate volume.

In conclusion the author takes great pleasure in expressing his appreciation of the valuable suggestions made by Professor F. S. Woods of the Massachusetts Institute of Technology, as well as of those by Professor L. G. Weld of the University of Iowa and Professor E. J. Townsend of the University of Illinois.

ABRAHAM COHEN.

JOHNS HOPKINS UNIVERSITY, BALTIMORE, MARYLAND,
October, 1906.

CONTENTS

CHAPTER I

Differential Equations and their Solutions

SECTION PAGE

1. Differential Equation. Ordinary and Partial. Order, Degree . . . 1
2. Solution of an Equation 2
3. Derivation of a Differential Equation from its Primitive . . . 3
4. General, Particular Solution 5

CHAPTER II

Differential Equations of the First Order and First Degree

5. Exact Differential Equation. Integrating Factor 7
6. General Plan of Solution 9
7. Condition that Equation be Exact 9
8. Exact Differential Equations 11
9. Variables Separated or Separable 13
10. Homogeneous Equations 14
11. Equations in which M and N are Linear but not Homogeneous . . 16
12. Equations of the Form $yf_1(xy)dx + xf_2(xy)dy = 0$ 17
13. Linear Equations of the First Order 18
14. Equations Reducible to Linear Equations 20
15. Equations of the Form $x^r y^s(my\,dx + nx\,dy) + x^\rho y^\sigma\,(\mu y\,dx + \nu x\,dy) = 0$. 22
16. Integrating Factors by Inspection 23
17. Other Forms for which Integrating Factors can be Found . . . 24
18. Transformation of Variables 26
19. Summary 28

CHAPTER III

Applications

20. Differential Equation of a Family of Curves 31
21. Geometrical Problems involving the Solution of Differential Equations 34

SECTION PAGE
22. Orthogonal Trajectories 38
23. Physical Problems giving Rise to Differential Equations . . . 43

CHAPTER IV

DIFFERENTIAL EQUATIONS OF THE FIRST ORDER AND HIGHER DEGREE THAN THE FIRST

24. Equations Solvable for p 49
25. Equations Solvable for y 52
26. Equations Solvable for x 55
27. Clairaut's Equation 56
28. Summary 58

CHAPTER V

SINGULAR SOLUTIONS

29. Envelopes 61
30. Singular Solutions 63
31. Discriminant 64
32. Singular Solution Obtained directly from the Differential Equation . 66
33. Extraneous Loci 69
34. Summary 75

CHAPTER VI

TOTAL DIFFERENTIAL EQUATIONS

35. Total Differential Equations 76
36. Method of Solution 80
37. Homogeneous Equations 81
38. Equations involving more than Three Variables 83
39. Equations which do not satisfy the Condition for Integrability . . 84
40. Geometrical Interpretation 86
41. Summary 87

CHAPTER VII

LINEAR DIFFERENTIAL EQUATIONS WITH CONSTANT COEFFICIENTS

42. General Linear Differential Equation 89
43. Linear Differential Equations with Constant Coefficients. Complementary Function 91

SECTION PAGE
44. Roots of Auxiliary Equation Repeated 93
45. Roots of the Auxiliary Equation Complex 94
46. Properties of the Symbolic Operator $(D-\alpha)$ 96
47. Particular Integral 97
48. Another Method of finding the Particular Integral 101
49. Variation of Parameters 103
50. Method of Undetermined Coefficients 107
51. Cauchy's Linear Equation 113
52. Summary 115

CHAPTER VIII

Linear Differential Equations of the Second Order

53. Change of Dependent Variable 123
54. Change of Independent Variable 127
55. Summary 129

CHAPTER IX

Miscellaneous Methods for Solving Equations of Higher Order than the First

56. General Plan of Solution 131
57. Dependent Variable Absent 131
58. Independent Variable Absent 134
59. Linear Equations with Particular Integral Known 135
60. Exact Equation. Integrating Factor 137
61. Transformation of Variables 143
62. Summary 144

CHAPTER X

Systems of Simultaneous Equations

63. General Method of Solution 149
64. Systems of Linear Equations with Constant Coefficients . . . 150
65. Systems of Equations of the First Order 153
66. Geometrical Interpretation 157
67. Systems of Total Differential Equations 159
68. Differential Equations of Higher Order than the First Reducible to Systems of Equations of the First Order 160
69. Summary 161

CHAPTER XI

INTEGRATION IN SERIES

SECTION PAGE

70. The Existence Theorem 164
71. Singular Solutions 168
72. Integration, in Series, of an Equation of the First Order . . . 169
73. Riccati's Equation 173
74. Integration, in Series, of Equations of Higher Order than the First . 177
75. Gauss's Equation. Hypergeometric Series 192

CHAPTER XII

PARTIAL DIFFERENTIAL EQUATIONS

76. Primitives involving Arbitrary Constants 196
77. Primitives involving Arbitrary Functions 199
78. Solution of a Partial Differential Equation 202

CHAPTER XIII

PARTIAL DIFFERENTIAL EQUATIONS OF THE FIRST ORDER

79. Linear Partial Differential Equations of the First Order. Method of Lagrange 205
80. Integrating Factors of the Ordinary Differential Equation $M\,dx+N\,dy=0$ 209
81. Non-linear Partial Differential Equations of First Order. Complete, General, Singular Solutions 211
82. Method of Lagrange and Charpit 215
83. Special Methods 221
84. Summary 227

CHAPTER XIV

PARTIAL DIFFERENTIAL EQUATIONS OF HIGHER ORDER THAN THE FIRST

85. Partial Differential Equations of the Second Order, Linear in the Second Derivatives. Monge's Method 230
86. Special Method 236
87. General Linear Partial Differential Equations 238
88. Homogeneous Linear Equations with Constant Coefficients . . . 239

SECTION PAGE
89. Roots of Auxiliary Equation Repeated 240
90. Roots of Auxiliary Equation Complex 242
91. Particular Integral 243
92. Non-homogeneous Linear Equations with Constant Coefficients . . 246
93. Equations Reducible to Linear Equations with Constant Coefficients . 249
94. Summary 251

NOTES

I. Condition that a Relation exist between Two Functions of Two Variables 253
II. General Summary 254

Answers to Examples 257
Index 269
References 271

DIFFERENTIAL EQUATIONS

CHAPTER I

DIFFERENTIAL EQUATIONS AND THEIR SOLUTIONS

1. Differential Equation. Ordinary and Partial. Order, Degree.—A *differential equation* is an equation involving differentials or derivatives. Thus,

(1) $$\frac{d^2y}{dx^2} + a^2y = 0,$$

(2) $$\left(x\frac{dy}{dx} - y\right)^2 = x^2 + y^2,$$

(3) $$y - x\frac{dy}{dx} + 3\frac{dx}{dy} = 0,$$

(4) $$\frac{\frac{d^2y}{dx^2}}{\left[1 + \left(\frac{dy}{dx}\right)^2\right]^{\frac{3}{2}}} - k = 0,$$

(5) $$x\frac{\partial z}{\partial x} + y\frac{\partial z}{\partial y} - z = 0,$$

(6) $$y\frac{\partial^2 z}{\partial x^2} + 2x\frac{\partial^2 z}{\partial x\,dy} - \frac{\partial z}{\partial y} = xyz,$$

(7) $$x^2y^3dx + x^3y^2dy = 0,$$

are examples of differential equations.

Equations in which there is a single independent variable (and which, therefore, involve ordinary derivatives) are known as *ordinary differential equations*. Equations (1), (2), (3), (4), (7) are such.

If an equation involves more than one independent variable, so that partial derivatives enter, it is known as a *partial differential equation*. Examples of such are equations (5), (6).

By the *order* of an equation we mean the order of the highest derivative involved. Thus, equations (2), (3), (5), (7), are of the first order; (1), (4), (6), are of the second order.

By the *degree* of an equation, we mean the degree of the highest ordered derivative entering, when the equation is rationalized and cleared of fractions with regard to all the derivatives. Thus (1), (5), (6), (7), are of the first degree; (2), (3), (4), are of the second degree.

2. Solution of an Equation. — By a *solution* of a differential equation we mean a relation connecting the dependent and independent variables which satisfies the equation. Thus $y = \sin ax$ is a solution of (1), $x^2 + y^2 = \frac{1}{k^2}$ is a solution of (4), $z = x + y$ is a solution of (5), $x^3y^3 = 1$ is a solution of (7). [The student, as an exercise, should verify these facts.]

Attention should be called to the fact that a differential equation has an indefinite number of solutions. It can be seen readily that $y = 2 \sin ax$, $y = 6 \cos ax$, $y = A \cos ax + B \sin ax$ (where A and B are any constants whatever) all satisfy equation (1). The student is in the habit of adding a constant of integration when integrating a function. He says the integral of $\cos x$ is $\sin x + c$. Now the problem of integration studied in the Calculus is only a special case of the general problem of solving a differential equation. To integrate $\int \cos x \, dx$ is to find a function, $\sin x + c$, whose derivative is $\cos x$. In the language of the Differential Equations, we should say that the solution of $\frac{dy}{dx} = \cos x$ is $y = \sin x + c$, where c is an arbitrary constant.*

A constant in a solution will be said to be *arbitrary* if any value whatever may be assigned to it. Thus $y = \sin x + c$ is a solution of $\frac{dy}{dx} = \cos x$, no matter

* It may be noted that in the special case occurring in the Integral Calculus the arbitrary constant always occurs as an *additive* one, while in the general case it may enter in an endless number of ways.

what value c has. Similarly $y = A\cos ax + B\sin ax$ is a solution of (1) for any values of A and B. They are arbitrary constants. On the other hand, a is not arbitrary. While any value one pleases may be assigned to it in (1), once chosen, its value is fixed, and that value alone can enter into the expression for the solution.

Restricting ourselves to ordinary differential equations,* we see that a solution may involve one or more arbitrary constants. The question naturally arises, what is the maximum number of such constants a solution may contain?

3. Derivation of a Differential Equation from its Primitive. — Just as the problem of integration is the inverse of that of differentiation, so the problem of finding the solution of a differential equation is the inverse of that of finding the differential equation which is satisfied by a relation among a set of variables, which relation may or may not involve one or more arbitrary constants. In order to make this problem precise, we shall say that we wish to find the differential equation of lowest order satisfied by this relation, and not involving any arbitrary constants. Thus $y = A\cos x$, where A is an arbitrary constant, satisfies $\frac{dy}{dx} + y\tan x = 0$, $\frac{d^2y}{dx^2} + y = 0$, $\frac{d^3y}{dx^3} - y\tan x = 0$, etc. But we shall say that $\frac{dy}{dx} + y\tan x = 0$ is *the* differential equation to which it gives rise.

Again, $y = A\cos x + B\sin x$, where A and B are arbitrary constants, satisfies $\frac{d^2y}{dx^2} + y = 0$, also $\frac{d^3y}{dx^3} + \frac{dy}{dx} = 0$, etc. Here, as before, $\frac{d^2y}{dx^2} + y = 0$ is *the* differential equation we are interested in.

Perfectly generally, if we have a relation which involves n arbitrary constants,† we differentiate this expression n times, thus having in

* The study of partial differential equations will be taken up in Chapter XII.

† It is implied, of course, that the n constants are *essential*; that is, that they cannot be replaced by a smaller number. For example, $y = x + a + b$ really involves only one essential constant, since $a + b$ is no more than a single constant. Again ae^{x+b} is no more general than ae^x.

all $n+1$ equations from which to eliminate the n constants. So that a relation (or, as we shall henceforth call it, a *primitive*) in which n arbitrary constants appear, gives rise to a differential equation which involves derivatives of as high an order as the nth. This process is unambiguous; hence a primitive gives rise to one and only one differential equation. Without going into a rigorous proof of the fact, we see how *a primitive involving n arbitrary constants gives rise to a differential equation of the nth order.*

To illustrate, find the differential equations corresponding to the following primitives: —

Ex. 1. $y = c_1e^{a_1x} + c_2e^{a_2x}$. Here c_1 and c_2 are the arbitrary constants.

Then $\frac{dy}{dx} = a_1c_1e^{a_1x} + a_2c_2e^{a_2x}$,

$$\frac{d^2y}{dx^2} = a_1^2c_1e^{a_1x} + a_2^2c_2e^{a_2x}.$$

From these three equations we must eliminate c_1 and c_2. Considering them as three homogeneous equations in the quantities 1, $c_1e^{a_1x}$, $c_2e^{a_2x}$, we have

$$\begin{vmatrix} y, & 1, & 1 \\ \frac{dy}{dx}, & a_1, & a_2 \\ \frac{d^2y}{dx^2}, & a_1^2, & a_2^2 \end{vmatrix} = 0, \text{ or } \frac{d^2y}{dx^2} - (a_1 + a_2)\frac{dy}{dx} + a_1a_2y = 0.$$

Ex. 2. $(x-c)^2 + y^2 = r^2$. Here c is the arbitrary constant. Then

$x - c + y\frac{dy}{dx} = 0$. From these two equations we must eliminate c.

Now $x - c = -y\frac{dy}{dx}$. Substituting this in the original equation, we have

$$y^2\left(\frac{dy}{dx}\right)^2 + y^2 = r^2.$$

Ex. 3. $y = cx + \sqrt{1 - c^2}$.

Ex. 4. $(x - c_1)^2 + (y - c_2)^2 = r^2$.

Ex. 5. $y = c_1x^2 + c_2$.

Ex. 6. $y^2 + c_1x = 0$.

Ex. 7. $x^2 = 2cy + c^2$.

4. General, Particular Solution. — If now we start with a differential equation, its solution involving the maximum number of arbitrary constants is nothing but the primitive which gives rise to the differential equation. That solution cannot contain more than n arbitrary constants, by the theorem in § 3. Besides, it must contain as many as n; otherwise it would be the primitive of a lower ordered equation.

The solution involving the maximum number of arbitrary constants is called the *general* (or *complete*) solution.* By means of the general existence theorem (§ 70), we can prove the following theorem:— *The general solution of an ordinary differential equation of the nth order is one that involves n arbitrary constants.*

Attention should be called to the fact that although the general solution may assume a variety of forms, all of these give the same relation among the variables, so that there is actually only one general solution; thus it is readily seen that $x^3y^3 = C$ is a solution of (7); so also is $\log x + \log y = C$, or $\log xy = C$. These, obviously, are all equivalent to saying that xy is constant. The uniqueness of the general solution is part of the existence theorem.

A solution which is derivable from the general solution by assigning fixed values to the arbitrary constants is called a *particular* solution. Thus, $y = \cos x$ and $y = \cos x - \sin x$ are particular solutions of $\frac{d^2y}{dx^2} + y = 0$.

* We shall see later that there may be solutions which are distinct from the general solution. In the general theory of Differential Equations the existence of a solution for every differential equation (under certain restrictions) is proved. The solution there obtained is the general solution referred to in the text.

As mentioned in § 2, the problem of the Differential Equations includes that of the Integral Calculus as a special case. Thus, in the latter the general problem is to solve

$$\frac{dy}{dx}=f(x).$$

This is only a special case of the problem of finding the solution of the differential equation of the first order involving two variables,

$$\frac{dy}{dx}=f(x, y)$$

where $f(x, y)$ may be a function of both the variables. We speak of *integrating or solving the equation*, in the general case, and at times refer to the simpler problem of the Integral Calculus as performing a *quadrature.*

A function of the independent variable will be said to be an *integral* of the equation if, on equating it to the dependent variable, we have a solution. We have a *general* or *particular* integral according as the resulting solution is general or particular.

While the problem of finding the differential equation corresponding to a given primitive is a direct one, and can be carried out according to a general plan, involving simply differentiation and elimination, that of finding the primitive or general solution of a given differential equation, like most inverse problems, cannot be solved by any general method.

In the following chapters we shall bring out, in as systematic a manner as possible, some of the classes of equations whose solutions can be found.

We shall understand that the problem of the Differential Equations is solved when we have reduced it to one of quadratures, that is, to a mere process of the Integral Calculus. While in the general theory of the Calculus it is proved that every function has an integral, it may not be possible to express it. In such cases we shall content ourselves by simply indicating this final process.

CHAPTER II

DIFFERENTIAL EQUATIONS OF THE FIRST ORDER AND THE FIRST DEGREE

5. Exact Differential Equation. Integrating Factor. The general type of an equation of the first order and degree is

$$(1) \qquad M\,dx + N\,dy = 0,$$

where M and N are functions of x and y.

Making use of the theorem that every differential equation has a general solution (§ 70), this equation has a solution containing one arbitrary constant. Solving for the constant, the solution has the form

$$(2) \qquad u\,(x, y) = C.$$

The differential equation having this primitive is obviously

$$\frac{\partial u}{\partial x}dx + \frac{\partial u}{\partial y}dy = 0.$$

Since this must be the same equation as (1), we must have the corresponding coefficients proportional, *i.e.*

$$\frac{\frac{\partial u}{\partial x}}{M} = \frac{\frac{\partial u}{\partial y}}{N}.$$

If we call this common ratio μ, (which is, at most, a function of x and y), we have

$$\frac{\partial u}{\partial x} = \mu M, \quad \frac{\partial u}{\partial y} = \mu N.$$

So that
$$\mu(M\,dx + N\,dy) = du.$$

We shall speak of an expression which is the differential of a function of one or more variables as an *exact differential.* Thus, $\mu(M\,dx + N\,dy)$ is such, since it is the differential of u. We shall further speak of a differential equation as an *exact differential equation,* if, when all the terms in it are brought to one side, that member is an exact differential. The above result can now be stated as follows: *Assuming the existence of the general solution of the differential equation* (1), *a factor* $\mu(x, y)$ *exists which, when introduced, will make the equation exact.*

This factor is known as an *integrating factor,* because, as we shall see (§ 8), when our equation is exact, its integration can be effected readily.

A differential equation of the first order and degree has an indefinite number of integrating factors.

Suppose μ to be an integrating factor. Then

$$\mu\,(M\,dx + N\,dy) \equiv du\,(x, y)$$

where the sign of identity $\equiv$ means that $\mu M = \frac{\partial u}{\partial x}$ and $\mu N = \frac{\partial u}{\partial y}$.

Now, if $\phi(u)$ is any continuous function of u, we have

$$\mu\phi(u)M\,dx + \mu\phi(u)N\,dy = \phi(u)\,\frac{\partial u}{\partial x}\,dx + \phi(u)\,\frac{\partial u}{\partial y}\,dy \equiv d\psi(u) = d\Psi(x, y),$$

where
$$\frac{d\psi(u)}{du} = \phi(u), \text{ or } \psi(u) = \int \phi(u)\,du.$$

Hence $\mu\phi(u)$ is also an integrating factor. Since $\phi(u)$ may be chosen in an indefinite number of ways, we see that the number of integrating factors is infinite. [For another proof, see Ex., § 7; also § 80.] Thus, it is obvious by inspection that $x\,dy - y\,dx = 0$ has $\frac{1}{x^2}$ for an integrating factor. We have, actually, $\frac{x\,dy - y\,dx}{x^2} \equiv d\left(\frac{y}{x}\right)$. Here $\mu = \frac{1}{x^2}$, $u = \frac{y}{x}$. Then $\frac{1}{x^2}\phi\left(\frac{y}{x}\right)$ will be an integrating factor. In particular, $\frac{1}{x^2}\frac{x}{y}$ or $\frac{1}{xy}$ is an integrating factor giving $\frac{dy}{y} - \frac{dx}{x}$ which is $d\left(\log\frac{y}{x}\right)$. Similarly, $\frac{1}{x^2}\frac{x^2}{y^2}$ or $\frac{1}{y^2}$ gives $d\left(-\frac{x}{y}\right)$.

6. General Plan of Solution. — Since every differential equation of the first order and degree which can be solved by elementary means has integrating factors, it would seem natural to try to find such a factor when the problem of solving an equation of this type arises. Practically, this is not always possible or desirable. In the following paragraphs of this chapter will be found the more important and the more frequently occurring classes of equations of the first order and degree which can be solved by elementary means; and it will be noticed that they will be solved, in general, either by finding integrating factors for them or by transforming them into other forms for which integrating factors are known.

7. Condition that Equation be Exact. — If the equation is exact to begin with, of course, no integrating factor need be sought. We must, then, find the necessary and sufficient condition for exactness of an equation. If

(1) $$M\,dx + N\,dy = 0$$

is exact, that is, if $M\,dx + N\,dy$ is the differential of some function u, then $\frac{\partial u}{\partial x} = M$ and $\frac{\partial u}{\partial y} = N$, and

(2) $$\frac{\partial M}{\partial y} = \frac{\partial N}{\partial x},$$

since $\frac{\partial^2 u}{\partial x\,\partial y} = \frac{\partial^2 u}{\partial y\,\partial x}$.* (2) is, then, a necessary condition for exactness of the equation.

We shall now prove that it is also sufficient. Even more, we shall show that if (2) holds, we can actually find a function u such that its differential is $M\,dx + N\,dy$, or, what is the same thing, such that

(3) $$\frac{\partial u}{\partial x} = M, \text{ and } \frac{\partial u}{\partial y} = N.$$

* Assuming the continuity of M and N, and the existence and continuity of $\frac{\partial M}{\partial y}$ and $\frac{\partial N}{\partial x}$.

In order that the first of these relations (3) should hold, we must have

$$u=\int^{x} M\,dx+Y(y)\,^{*} \tag{4}$$

where Y is a function of y only, and plays the rôle of a constant of integration, since y is considered a constant in the process of integration involved in (4). The value of u given by (4) will satisfy the first of (3), no matter what function of y Y may be. In order that u satisfy the second of (3) we must have

$$\frac{\partial u}{\partial y}=\frac{\partial}{\partial y}\int^{x} M\,dx+\frac{dY}{dy}=N;$$

that is, Y must satisfy the equation

$$\frac{dY}{dy}=N-\frac{\partial}{\partial y}\int^{x} M\,dx. \tag{5}$$

Since the left-hand member of (5) is a function of y only, the same must be true of the right-hand member; that is, the latter must be free of x, or in other words it must be a constant as far as x is concerned, and its derivative with respect to x must be zero. As a matter of fact, that derivative is $\dfrac{\partial N}{\partial x}-\dfrac{\partial M}{\partial y}$,† which is zero because of (2). We can, then, find Y to satisfy (5), viz.

$$Y=\int\left[N-\frac{\partial}{\partial y}\int^{x} M\,dx\right]dy,$$

and the resulting value of u in (4) will satisfy both equations (3).

Ex. Using the fact that (2) is the necessary and sufficient condition for exactness, prove that if μ is an integrating factor, such that $\mu(M\,dx+N\,dy)\equiv du$, $\mu\,\phi(u)$ is also an integrating factor.

* By $\int^{x} M\,dx$ we mean the result of integrating $M\,dx$ considering y as a constant.

† Obviously $\dfrac{\partial}{\partial x}\int^{x} M\,dx=M$, since, in both of the processes involved, y is considered constant.

8. Exact Differential Equations.—This suggests a method of solving an exact equation. For, in this case, $M\,dx + N\,dy = du$ where

$$u = \int^x M\,dx + \int\left[N - \frac{\partial}{\partial y}\int^x M\,dx\right]dy.$$

So that the general solution is

$$\int^x M\,dx + \int\left[N - \frac{\partial}{\partial y}\int^x M\,dx\right]dy = c. \tag{6}$$

Expressed in words, the operations involved in (6) are: *Integrate $M\,dx$, considering y as a constant, thus obtaining $\int^x M\,dx$. Subtracting the derivative of this, with respect to y, from N, a function of y only is left. The integral of this function plus $\int^x M\,dx$ is the left-hand member of* (6).

Remark.—Frequently $N - \frac{\partial}{\partial y}\int^x M\,dx$ is nothing but those terms of N free of x. This suggests the simple rule: *Integrate $M\,dx$, considering y as a constant; then integrate those terms in $N\,dy$ free of x. The sum of these equated to an arbitrary constant is the general solution.* This rule may fail because $\int^x M\,dx$ is not unique as far as terms involving y only are concerned. Thus $\int^x (x+y)dx$ may be either $\frac{1}{2}x^2 + xy$ or $\frac{1}{2}(x+y)^2$. See also Ex. 3. But the rule is found to work so often that it seems worth mentioning, with the understanding, however, that when it is employed, the result be redifferentiated to see whether the original equation is obtained.

As an exercise let the student show that the general solution may also be obtained in the form

$$\int^y N\,dy + \int\left[M - \frac{\partial}{\partial x}\int^y N\,dy\right]dx = c.$$

Ex. 1. $\dfrac{2xy+1}{y}dx + \dfrac{y-x}{y^2}dy = 0.$

$$\frac{\partial}{\partial y}\left(\frac{2xy+1}{y}\right) = -\frac{1}{y^2} = \frac{\partial}{\partial x}\left(\frac{y-x}{y^2}\right);$$ hence equation is exact.

$$\int^x \frac{2xy+1}{y}dx = x^2 + \frac{x}{y}.$$ $\frac{1}{y}$ is the only term in N free of x.

$$\int \frac{1}{y}\,dy = \log y.$$

$\therefore\ x^2 + \dfrac{x}{y} + \log y = c$ is the general solution.

Ex. 2. $\dfrac{y^2 - 2x^2}{xy^2 - x^3}dx + \dfrac{2y^2 - x^2}{y^3 - x^2y}dy = 0.$

Let the student prove that this is exact by showing that

$$\frac{\partial M}{\partial y} = \frac{\partial N}{\partial x}.$$

$$\int^x \frac{y^2 - 2x^2}{xy^2 - x^3}dx = \int^x \frac{(y^2 - x^2) - x^2}{x(y^2 - x^2)}dx = \int \frac{dx}{x} - \int^x \frac{x}{y^2 - x^2}dx =$$

$$\log x + \frac{1}{2}\log(y^2 - x^2).$$

At first sight it might seem that no term of N is free of x. But writing it in the form $\dfrac{y^2 + (y^2 - x^2)}{y(y^2 - x^2)} = \dfrac{y}{y^2 - x^2} + \dfrac{1}{y}$, we see that there is such a term, viz. $\dfrac{1}{y}$.

$\therefore$ the solution is $\log x + \frac{1}{2}\log(y^2 - x^2) + \log y = c$,

or $$x^2y^2(y^2 - x^2) = c.$$

Ex. 3. $\dfrac{dx}{\sqrt{x^2 + y^2}} + \left(\dfrac{1}{y} - \dfrac{x}{y\sqrt{x^2 + y^2}}\right)dy = 0.$

Let the student prove that this is exact.

$\displaystyle\int^x \frac{dx}{\sqrt{x^2 + y^2}} = \log(x + \sqrt{x^2 + y^2})$, the form given in integral tables.

N contains the term $\dfrac{1}{y}$ which is free of x. Hence if the rule is followed blindly, the solution would seem to be

$$\log(x + \sqrt{x^2 + y^2}) + \log y = c.$$

As a matter of fact, the solution is $\log(x+\sqrt{x^2+y^2})=c$.

The trouble here is with the value of $\int^x \frac{dx}{\sqrt{x^2+y^2}}$.

$$\int\frac{dx}{\sqrt{x^2+a^2}}=\int\frac{d\frac{x}{a}}{\sqrt{\left(\frac{x}{a}\right)^2+1}}=\log\left(\frac{x}{a}+\sqrt{\left(\frac{x}{a}\right)^2+1}\right)$$

$$=\log(x+\sqrt{x^2+a^2})-\log a.$$

Naturally in tables of integrals the term $-\log a$ is omitted, as that may be incorporated in the constant of integration. In the case of the problem under discussion it makes a big difference whether this term is used or not. If it is used, the rule gives the correct result. But if the form of the integral as given in the tables is used, then the rule is at fault. Hence the caution given in the remark above should be heeded.

Ex. 4. $(y+x)dx+x\,dy=0$.

Ex. 5. $(6x-2y+1)dx+(2y-2x-3)dy=0$.

9. Variables Separated or Separable. In case M is a function of x only, and N one of y only, the relation $\frac{\partial M}{\partial y}=\frac{\partial N}{\partial x}$ is obviously satisfied. In this case we shall say the *variables are separated.* The integral is, of course,

$$\int M\,dx+\int N\,dy=c.$$

Very frequently, the variables are separable by inspection.

Ex. 1. $\sec x\cos^2 y\,dx-\cos x\sin y\,dy=0$.

Hence $$\frac{\sec x}{\cos x}dx-\frac{\sin y}{\cos^2 y}dy=0,\text{*}$$

or $$\sec^2 x\,dx-\tan y\sec y\,dy=0.$$

The general solution is $\tan x-\sec y=c$.

* Here $\frac{1}{\cos x\cos^2 y}$ is an integrating factor, found by inspection.

Ex. 2. $(1+x)y^2\,dx - x^3\,dy = 0.$

Ex. 3. $2(1-y^2)xy\,dx + (1+x^2)(1+y^2)\,dy = 0.$

Ex. 4. $\sin x \cos^2 y\,dx + \cos^2 x\,dy = 0.$

10. Homogeneous Equations. A widely occurring class of equations where the variables can be separated, not by inspection, but by a simple transformation of variables, is that in which M and N are homogeneous functions of x and y, and of the same degree. Then $\frac{M}{N}$ is a homogeneous function of degree zero, and is therefore a function of $\frac{y}{x}$.*

Our equation can now be written

$$\frac{dy}{dx} = -\frac{M}{N} = F\left(\frac{y}{x}\right).$$

Let $\frac{y}{x}$ be a new variable, say v.

Then $y = vx$, $$\frac{dy}{dx} = v + x\frac{dv}{dx} = F(v),$$

or $$\frac{dv}{F(v) - v} = \frac{dx}{x},$$

and our variables are separated.

Integrating this, and then replacing v by its value in terms of x and y, we have the general solution.

* A convenient definition of a *homogeneous function* of x and y of degree r is, that if in the function we replace x and y by tx and ty respectively, where t is anything we please, the result will be the original function multiplied by t^r. (This definition can be extended at once to a function of any number of variables. It is obviously consistent with the old definition of homogeneity of polynominals.) This definition can be formulated thus: If $f(x, y)$ is a homogeneous function of degree r, then $f(tx, ty) = t^r f(x, y)$. If we put $t = \frac{1}{x}$, we have $f\left(1, \frac{y}{x}\right) = \frac{1}{x^r} f(x, y)$, or $f(x, y) = x^r f\left(1, \frac{y}{x}\right)$. When $r = 0$, we have $f(x, y) = f\left(1, \frac{y}{x}\right) = F\left(\frac{y}{x}\right)$.

Ex. 1. $\left(xe^{\frac{y}{x}}+y\right)dx - x\,dy = 0.$

Put $y = vx$, $dy = v\,dx + x\,dv$. Then

$$x(e^v + v)\,dx - xv\,dx - x^2\,dv = 0,$$

or $$\frac{dx}{x} - \frac{dv}{e^v} = 0. \quad \text{Integrating, we have}$$

$$\log x + e^{-v} = c, \text{ or } \log x + e^{-\frac{y}{x}} = c.$$

Ex. 2. $2x^2y + 3y^3 - (x^3 + 2xy^2)\frac{dy}{dx} = 0.$

Put $y = vx$, $\frac{dy}{dx} = v + x\frac{dv}{dx}$. Then

$$\frac{1 + 2v^2}{v + v^3}\,dv - \frac{dx}{x} = 0,$$

or $$\frac{dv}{v} + \frac{v\,dv}{1+v^2} - \frac{dx}{x} = 0. \quad \text{Integrating, we have}$$

$$\log v + \tfrac{1}{2}\log(1 + v^2) - \log x = k,$$

or $$\log v^2 + \log(1 + v^2) - \log x^2 = 2k,$$

or $$\frac{v^2(1 + v^2)}{x^2} = e^{2k} = c; \text{ whence}$$

$$y^2(x^2 + y^2) = cx^6.$$

Ex. 3. $(y^2 - xy)\,dx + x^2\,dy = 0.$

Ex. 4. $2x^2y + y^3 - x^3\frac{dy}{dx} = 0.$

Ex. 5. $y^3dx + x^3dy = 0.$

Ex. 6. $\left(x + y\cos\frac{y}{x}\right)dx - x\cos\frac{y}{x}\,dy = 0.$

11. Equations in which *M* and *N* are Linear but not Homogeneous. If M and N are both of the first degree but are not homogeneous, we can, by a very simple transformation, make them homogeneous.

Suppose our equation to be of the form

$$(a_1x + b_1y + c_1)\,dx + (a_2x + b_2y + c_2)\,dy = 0.$$

Put $x = x' + \alpha$, $y = y' + \beta$. The equation becomes

$$(a_1x' + b_1y' + a_1\alpha + b_1\beta + c_1)\,dx' + (a_2x' + b_2y' + a_2\alpha + b_2\beta + c_2)\,dy' = 0.$$

Now we may choose α and β such that

$$a_1\alpha + b_1\beta + c_1 = 0,$$

and

$$a_2\alpha + b_2\beta + c_2 = 0;$$

i.e. if we put $\quad \alpha = \dfrac{b_1c_2 - b_2c_1}{a_1b_2 - a_2b_1}$, and $\beta = \dfrac{c_1a_2 - c_2a_1}{a_1b_2 - a_2b_1}$,

our equation takes the form

$$(a_1x' + b_1y')\,dx' + (a_2x' + b_2y')\,dy' = 0,$$

where the coefficients are homogeneous.

Ex. 1. $(4x + 3y + 1)\,dx + (x + y + 1)\,dy = 0.$

Putting $x = x' + 2$, $y = y' - 3$, this becomes

$$(4x' + 3y')\,dx' + (x' + y')\,dy' = 0.$$

Now, if $y' = vx'$, we get

$$\frac{1 + v}{4 + 4v + v^2}\,dv + \frac{dx'}{x'} = 0,$$

or

$$\frac{dv}{2 + v} - \frac{dv}{(2 + v)^2} + \frac{dx'}{x'} = 0.$$

$$\therefore \log(2 + v) + \frac{1}{2 + v} + \log x' = c,$$

or

$$\log x'(2 + v) = c - \frac{1}{2 + v}, \text{ whence,}$$

$$\log(2x' + y') = c - \frac{x'}{2x' + y'}.$$

Passing back to x and y, this becomes

$$\log(2x+y-1) = c - \frac{x-2}{2x+y-1}.$$

Ex. 2. $(4x-y+2)\,dx + (x+y+3)\,dy = 0.$

Remark.—This method breaks down in case $a_1 b_2 - a_2 b_1 = 0$. But in this case we can find a transformation which will separate the variables at once. For we have $\frac{a_2}{a_1} = \frac{b_2}{b_1} = k$, a constant, and the equation takes the form

$$(a_1 x + b_1 y + c_1)\,dx + [k(a_1 x + b_1 y) + c_2]\,dy = 0.$$

If now, we put $a_1 x + b_1 y = t$, so that $y = \frac{t - a_1 x}{b_1}$, our equation takes the form

$$(t + c_1)\,dx + (kt + c_2)\left(\frac{dt - a_1\,dx}{b_1}\right) = 0,$$

or

$$dx + \frac{kt + c_2}{(b_1 - a_1 k)\,t + b_1 c_1 - a_1 c_2}\,dt = 0,$$

where the variables are separated.

Ex. 3. $(2x+y)\,dx - (4x+2y-1)\,dy = 0.$

12. Equations of the Form $yf_1(xy)\,dx + xf_2(xy)\,dy = 0$. Another class of equations in which the variables can be separated by a simple transformation may be mentioned. The general type of such an equation is

$$yf_1(xy)\,dx + xf_2(xy)\,dy = 0,$$

where by $f(xy)$ we mean a function of the product xy.

If we put $xy = v$, or $y = \frac{v}{x}$, then $x\,dy = dv - \frac{v}{x}\,dx$, and our equation becomes

$$\frac{v}{x} f_1(v)\,dx + f_2(v)\,dv - \frac{v}{x} f_2(v)\,dx = 0,$$

or

$$\frac{f_2\,dv}{v(f_1 - f_2)} + \frac{dx}{x} = 0,$$

where the variables are separated.

Ex. 1. $(y + 2xy^2 - x^2y^3)\,dx + 2x^2y\,dy = 0.$

Putting $y = \dfrac{v}{x}$, $dy = \dfrac{x\,dv - v\,dx}{x^2}$, our equation becomes

$$\frac{v}{x}(1 + 2v - v^2)\,dx + 2xv\frac{x\,dv - v\,dx}{x^2} = 0,$$

or

$$\frac{2\,dv}{1 - v^2} + \frac{dx}{x} = 0.$$

Integrating, we get, since $\dfrac{2}{1 - v^2} = \dfrac{1}{1 + v} + \dfrac{1}{1 - v}$,

$$\log\frac{1 + v}{1 - v} + \log x = k,$$

or

$$x\,\frac{1 + v}{1 - v} = c.$$

Replacing v by its value, we get

$$\frac{x + x^2y}{1 - xy} = c.$$

Ex. 2. $(2y + 3xy^2)\,dx + (x + 2x^2y)\,dy = 0.$

Ex. 3. $(y + xy^2)\,dx + (x - x^2y)\,dy = 0.$

13. Linear Equations of the First Order. A linear differential equation (of any order) is a special kind of equation of the first degree. It is not only of the first degree in the highest derivative, but it is of the first degree in the dependent variable and all its derivatives. Thus, the general type of a *linear differential equation of the first order* is

$$\frac{dy}{dx} + Py = Q$$

where P and Q are any functions of x only.

An integrating factor for this equation is readily seen* to be $e^{\int P dx}$, since $\frac{d}{dx}(ye^{\int P dx}) = e^{\int P dx}\left(\frac{dy}{dx} + Py\right)$. Introducing this factor, we have the solution by means of a single quadrature, thus

$$ye^{\int P dx} = \int Q e^{\int P dx}\,dx + c.\dagger$$

Ex. 1. $\frac{dy}{dx} + y \cot x = \sec x.$

Here $e^{\int P dx} = e^{\int \cot x dx} = e^{\log \sin x} = \sin x$ is an integrating factor. Using this we have

$$y \sin x = \int \sin x \sec x \, dx = \int \tan x \, dx = -\log \cos x + c.$$

Hence the solution is $\qquad y \sin x = -\log \cos x + c,$

or $\qquad y \sin x = \log \sec x + c.$

Ex. 2. $x\frac{dy}{dx} + (1 + x)y = e^x.$

* We shall see later (§ 80) that this form of integrating factor arises by a perfectly general method. Excepting for pedagogic reasons, that method could be given at this point, without assuming anything to prevent our using it in this connection.

† While the above method of solving a linear differential equation of the first order is undoubtedly the simplest in both theory and practice, and besides has the advantage of being readily retained in mind, it may be well to call attention to a second method, which is part of a general method applicable to linear equations of any order (see §§ 53, 59).

Let $y = y_1 v$. Then $\frac{dy}{dx} = y_1 \frac{dv}{dx} + \frac{dy_1}{dx} \cdot v$; and the equation becomes $y_1 \frac{dv}{dx} + \left(\frac{dy_1}{dx} + Py_1\right) v = Q$. If we choose y_1 so that $\frac{dy_1}{dx} + Py_1 = 0$, we get $y_1 = e^{-\int P dx}$, and the equation in v becomes $\frac{dv}{dx} = Q e^{\int P dx}$, whence $v = \int Q e^{\int P dx}\,dx + c$; and we have finally $y = e^{-\int P dx} \int Q e^{\int P dx}\,dx + ce^{-\int P dx}$.

Putting this in the proper form (where the coefficient of $\frac{dy}{dx}$ is unity), we get

$$\frac{dy}{dx}+\left(\frac{1}{x}+1\right)y=\frac{e^x}{x}.$$

$$\int P\,dx=\int\left(\frac{1}{x}\,dx+dx\right)=\log x+x.$$

∴ an integrating factor is $e^{\log x+x}$, or xe^x.

Using this, we get $yxe^x=\int e^{2x}\,dx=\frac{1}{2}e^{2x}+c.$

Ex. 3. $\frac{dy}{dx}-\frac{2y}{x+1}=(x+1)^3.$

Ex. 4. $(x+x^3)\frac{dy}{dx}+4x^2y=2.$

Ex. 5. $x^2\frac{dy}{dx}+(1-2x)y=x^2.$

14. Equations Reducible to Linear Equations. — At times an obvious transformation will change an equation into a linear one. Such a one, of frequent occurrence, is

$$\frac{dy}{dx}+Py=Qy^n\,*$$

where, as before, P and Q are functions of x only, and n is any number.

Dividing by y^n we get

$$y^{-n}\frac{dy}{dx}+Py^{-n+1}=Q.$$

Let $y^{-n+1}=v$, then $y^{-n}\frac{dy}{dx}=\frac{1}{1-n}\frac{dv}{dx}$, and our equation becomes

$$\frac{dv}{dx}+(1-n)Pv=(1-n)Q,$$

which is linear.

* This is known as Bernoulli's Equation, after James Bernoulli (1654–1705).

Ex. 1. $(1-x^3)\dfrac{dy}{dx} - 2(1+x)y = y^{\frac{5}{2}}$.

$$y^{-\frac{5}{2}}\frac{dy}{dx} - 2\frac{1+x}{1-x^3}y^{-\frac{3}{2}} = \frac{1}{1-x^3}.$$

Let $y^{-\frac{3}{2}} = v$, then $-\dfrac{3}{2}y^{-\frac{5}{2}}\dfrac{dy}{dx} = \dfrac{dv}{dx}$.

$$\therefore \frac{dv}{dx} + 3\frac{1+x}{1-x^3}v = -\frac{3}{2}\frac{1}{1-x^3}.$$

$$\int\frac{3+3x}{1-x^3}dx = 2\int\frac{dx}{1-x} + \int\frac{1+2x}{1+x+x^2}dx$$

$$= -2\log(1-x) + \log(1+x+x^2).$$

$\therefore$ an integrating factor is $e^{\log\frac{1+x+x^2}{(1-x)^2}}$, or $\dfrac{1+x+x^2}{(1-x)^2}$,

and we have $$v\frac{1+x+x^2}{(1-x)^2} = -\frac{3}{2}\int\frac{1}{1-x^3}\cdot\frac{1+x+x^2}{(1-x)^2}dx$$

$$= -\frac{3}{2}\int\frac{1}{(1-x)^3}dx = -\frac{3}{4}\frac{1}{(1-x)^2} + c;$$

whence $$y^{-\frac{3}{2}}\frac{1+x+x^2}{(1-x)^2} = -\frac{3}{4}\frac{1}{(1-x)^2} + c,$$

or $$y^{-\frac{3}{2}} = -\frac{3}{4}\frac{1}{1+x+x^2} + \frac{c(1-x)^2}{1+x+x^2}.$$

As examples of other cases where an obvious substitution will transform the equation into a linear one, consider the following:

Ex. 2. $y\dfrac{dy}{dx} + xy^2 = x$. [Put $y^2 = v$.]

Ex. 3. $\sin y\dfrac{dy}{dx} + \sin x\cos y = \sin x$. [Put $\cos y = v$.]

Ex. 4. $4x\dfrac{dy}{dx} + 3y + e^x x^4 y^5 = 0$.

Ex. 5. $\dfrac{dy}{dx} - \dfrac{y+1}{x+1} = \sqrt{y+1}$.

15. Equations of the Form $x^r y^s(my\,dx + nx\,dy) + x^\rho y^\sigma(\mu y\,dx + \nu x\,dy) = 0$.

* Since $d(x^a y^b) = x^{a-1}y^{b-1}(ay\,dx + bx\,dy)$, it is easily seen that if we start with any expression $x^r y^s(my\,dx + nx\,dy)$, we can make it exact by multiplying it by $x^\alpha y^\beta$, provided

$$cm = \alpha + r + 1, \text{ and } cn = \beta + s + 1,$$

where c is any number. As a matter of fact

$$x^{cm-r-1}y^{cn-s-1}x^r y^s(my\,dx + nx\,dy) = x^{cm-1}y^{cn-1}(my\,dx + nx\,dy)$$
$$= \frac{1}{c}d(x^{cm}y^{cn}).$$

If $c = 0$, this term must be replaced by $d \log x^m y^n$.

The object of introducing the undetermined quantity c is to enable us to find an integrating factor for an equation of the form

$$x^r y^s(my\,dx + nx\,dy) + x^\rho y^\sigma(\mu y\,dx + \nu x\,dy) = 0.$$

Just as the factor $x^{cm-r-1}y^{cn-s-1}$ will render the first set of terms exact, so $x^{\gamma\mu-\rho-1}y^{\gamma\nu-\sigma-1}$ will render the second set of terms exact. In order to find an integrating factor for the equation, we must so determine c and γ that these two factors are one and the same, *i.e.* we must have

$$cm - r - 1 = \gamma\mu - \rho - 1,$$
$$cn - s - 1 = \gamma\nu - \sigma - 1.$$

These two equations are sufficient, in general, to determine c and γ.*

Ex. 1. $x^4 y(3y\,dx + 2x\,dy) + x^2(4y\,dx + 3x\,dy) = 0.$

Here, $m = 3,\ n = 2,\ r = 4,\ s = 1,\ \mu = 4,\ \nu = 3,\ \rho = 2,\ \sigma = 0.$

$$3c - 5 = 4\gamma - 3, \qquad 2c - 2 = 3\gamma - 1.$$

From these we have, $c = 2$, $\gamma = 1$. Hence the integrating factor is xy^2. Introducing this, we get the solution

$$\tfrac{1}{2}x^6y^4 + x^4y^3 = c_1, \text{ or } x^6y^4 + 2x^4y^3 = k.$$

* If $m\nu = n\mu$, the equation reduces to the simple form $my\,dx - nx\,dy = 0$.

* $\frac{1}{c}d(x^{ac}y^{bc}) = x^{ac-1}y^{bc-1}(ay\,dx + bx\,dy)$

Ex. 2. $y^2(3y\,dx - 6x\,dy) - x(y\,dx - 2x\,dy) = 0.$

Ex. 3. $(2x^3y - y^2)\,dx - (2x^4 + xy)\,dy = 0.$

16. Integrating Factors by Inspection. — Integrating factors can frequently be found by inspection on closely examining the terms entering in the equation.* Of course no general rule for this can be given. A commonly occurring group of terms is $x\,dy - y\,dx$. This suggests $\dfrac{x\,dy - y\,dx}{x^2}$, or $\dfrac{x\,dy - y\,dx}{y^2}$, or $\dfrac{dy}{y} - \dfrac{dx}{x}$, or $\dfrac{x\,dy - y\,dx}{x^2\left(1 \pm \dfrac{y^2}{x^2}\right)}$, the factors being respectively $\dfrac{1}{x^2}, \dfrac{1}{y^2}, \dfrac{1}{xy}, \dfrac{1}{x^2 \pm y^2}$. So that $\dfrac{1}{x^2}$ is an integrating factor for an expression of the form $x\,dy - y\,dx + f(x)\,dx$, while $\dfrac{1}{y^2}$ may be used for one of the form $x\,dy - y\,dx + f(y)\,dy$, and $\dfrac{1}{xy}$ for $x\,dy - y\,dx + f(xy)(x\,dy + y\,dx)$, and $\dfrac{1}{x^2 \pm y^2}$ for $x\,dy - y\,dx + f(x^2 \pm y^2)(x\,dx \pm y\,dy)$. Other combinations will occur to one in actual practice.

Ex. 1. $(y^2 - xy)\,dx + x^2\,dy = 0.$ (Ex. 3, § 10.)

Writing this $\quad y^2\,dx - x(y\,dx - x\,dy) = 0,$

we see that $\dfrac{1}{xy^2}$ is an integrating factor. Using this, we get

$$\frac{dx}{x} - \frac{y\,dx - x\,dy}{y^2} = 0;$$

whence

$$\log x - \frac{x}{y} = c.$$

Ex. 2. $\dfrac{x\,dy - y\,dx}{\sqrt{x^2 - y^2}} = x\,dy.$

* An illustration of this we had in § 9, where by the introduction of a factor the variables were separated and the equation thus rendered exact.

Writing this $\dfrac{x\,dy - y\,dx}{x\sqrt{1-\left(\dfrac{y}{x}\right)^2}} = x\,dy$, we see at once that $\dfrac{1}{x}$ is an integrating factor. Using this, we get $\dfrac{x\,dy - y\,dx}{x^2\sqrt{1-\left(\dfrac{y}{x}\right)^2}} = dy$; whence

$$\sin^{-1}\frac{y}{x} = y + c.$$

Ex. 3. $(x+y)\,dx - (x-y)\,dy = 0.$

Writing this $x\,dx + y\,dy + y\,dx - x\,dy = 0$, we see at once that $\dfrac{1}{x^2+y^2}$ is an integrating factor.

Ex. 4. $(x^2+y^2)\,dx - 2\,xy\,dy = 0.$

Ex. 5. $(x-y^2)\,dx + 2\,xy\,dy = 0.$

Ex. 6. $x\,dy - y\,dx = (x^2+y^2)\,dx.$

17. Other Forms for which Integrating Factors can be Found.—In applying the test for exactness (§ 7), we find the value of $\dfrac{\partial M}{\partial y} - \dfrac{\partial N}{\partial x}$. If this turns out to be zero, the equation is exact. If not, it may contain either M or N as a factor. By the general method of § 80 (already referred to in a footnote, § 13), it will be seen that if $\dfrac{\dfrac{\partial M}{\partial y} - \dfrac{\partial N}{\partial x}}{N}$ is a function of x only, say $f_1(x)$, then $e^{\int f_1(x)dx}$ is an integrating factor, and if $\dfrac{\dfrac{\partial N}{\partial x} - \dfrac{\partial M}{\partial y}}{M}$ is a function of y only, say $f_2(y)$, then $e^{\int f_2(y)dy}$ is an integrating factor.

It may be interesting to call attention to the fact that this method of § 80 also informs us that $\frac{1}{xM + yN}$ is an integrating factor in case M and N are homogeneous and of the same degree, and that $\frac{1}{xM - yN}$ is an integrating factor if $M = yf_1(xy)$, $N = xf_2(xy)$.*

These two classes of equations were considered in §§ 10 and 12, where we found transformations that separated the variables. Leibnitz (1646–1716) and his school endeavored to solve all equations of the first order and degree by introducing new variables which become separable in the transformed equation. Euler (1707–1783) and his school tried to solve all equations of the first order and degree by finding integrating factors. As a matter of fact, these are frequently spoken of as *Euler factors or multipliers*. But the idea of an integrating factor seems to be due to a contemporary of his, Clairaut (1713–1765). Now it is interesting to note that just as every differential equation of the first order and degree has an integrating factor, in general, so it can be proved † that by a proper change of variables every such equation can be transformed into one in which the variables are separable. But in actual practice it would be awkward and difficult to carry out this method in all cases, just as it would be inadvisable to insist upon finding an integrating factor in every instance.

These two classes of equations are of particular interest as affording examples of cases where both the general methods of solution can be readily applied.

Ex. 1. $(3x^2 + 6xy + 3y^2)\,dx + (2x^2 + 3xy)\,dy = 0.$

$$\frac{\frac{\partial M}{\partial y} - \frac{\partial N}{\partial x}}{N} = \frac{1}{x}. \quad \therefore e^{\int \frac{dx}{x}} = x \text{ is an integrating factor.}$$

Ex. 2. $2\,xdx + (x^2 + y^2 + 2y)\,dy = 0.$

$$\frac{\frac{\partial N}{\partial x} - \frac{\partial M}{\partial y}}{M} = 1. \quad \therefore e^{\int dy} = e^y \text{ is an integrating factor.}$$

* These methods cease to apply if $xM + yN = 0$ in the first case, and if $xM - yN = 0$ in the second. But in either of these cases, the solution of the equation is effected directly with ease. For if $xM + yN = 0$, the equation takes the form $\frac{dy}{dx} = -\frac{M}{N} = \frac{y}{x}$, and its solution is $\frac{y}{x} = c$; on the other hand, when $xM - yN = 0$, the equation takes the form $\frac{dy}{dx} = -\frac{M}{N} = -\frac{y}{x}$, and the solution is $xy = c$.

† See Lie, *Differentialgleichungen*, Chapter 6, § 5; also the author's *Lie Theory*, § 20.

Ex. 3. $(y^4 + 2y)\,dx + (xy^3 + 2y^4 - 4x)\,dy = 0.$

Ex. 4. $(x^3y - y^4)\,dx + (y^3x - x^4)\,dy = 0.$

Ex. 5. Solve examples of §§ 10 and 12 by the method of this paragraph.

Ex. 6. $(y^2 - x^2 + 2\,mxy)dx + (my^2 - mx^2 - 2\,xy)\,dy = 0.$

18. Transformation of Variables. — In case the equation to be integrated does not come under any of the heads treated in this chapter, it is possible, at times, to reduce it to one of them by a transformation. No general rule for doing this can be formulated. The form of the equation must suggest the transformation to be tried. The following examples will illustrate.

Ex. 1. $x\,dy - y\,dx + 2\,x^2y\,dx - x^3\,dx = 0.$

Here $x\,dy - y\,dx$ suggests the transformation $\frac{y}{x} = v$. Making this transformation, our equation reduces to

$$\frac{dv}{dx} + 2\,xv = x, \text{ which is linear.}$$

An integrating factor is $e^{\int 2x dx}$ or e^{x^2}.

$$\therefore\ ve^{x^2} = \int xe^{x^2}\,dx = \frac{1}{2}e^{x^2} + c.$$

Hence the general solution is

$$\frac{y}{x}e^{x^2} = \frac{1}{2}e^{x^2} + c.$$

Ex. 2. $(x + y)\,dy - dx = 0.$

Here $x + y$ suggests the transformation $x + y = v$. Making this, our equation reduces to

$$v\,dv - (v + 1)dx = 0,$$

in which the variables are separable at once, so that we have

$$dx = \frac{v\,dv}{v + 1} = dv - \frac{dv}{v + 1}.$$

Integrating, we have $x = v - \log(v + 1) + c$,

i.e. $$x = x + y - \log(x + y + 1) + c,$$

or $$\log(x + y + 1) = y + c.$$

This can also be integrated as follows:

Writing it in the form

$$\frac{dx}{dy} - x = y,$$

we see it is linear, considering y as the independent variable.

An integrating factor is $$e^{-\int dy} = e^{-y}.$$

Hence $$xe^{-y} = \int ye^{-y}dy = -ye^{-y} - e^{-y} + c,$$

or $$x + y + 1 = ce^{y}.$$

Ex. 3. $x\,dx + y\,dy + y\,dx - x\,dy = 0$. (Ex. 3, § 16.)

Here $x\,dx + y\,dy$ suggests $x^2 + y^2$, while $y\,dx - x\,dy$ suggests $\frac{y}{x}$. This combination suggests the transformation $x^2 + y^2 = r^2$, $\frac{y}{x} = \tan\theta$; or, what is the same thing, $x = r\cos\theta$, $y = r\sin\theta$. Then

$$x\,dx + y\,dy = r\,dr,$$

$$y\,dx - x\,dy = -r^2\,d\theta,$$

$$dx = \cos\theta\,dr - r\sin\theta\,d\theta,$$

$$dy = \sin\theta\,dr + r\cos\theta\,d\theta.$$

Our equation then becomes

$$\frac{dr}{r} - d\theta = 0,$$

and the solution is

$$\log r - \theta = c, \text{ or } \log\sqrt{x^2 + y^2} - \tan^{-1}\frac{y}{x} = c.$$

Ex. 4. $x\dfrac{dy}{dx} - ay + by^2 = cx^{2a}$.* This special form is characterized by the fact that, when the first term is $x\dfrac{dy}{dx}$, the coefficient of y is the negative of half the exponent of x in the right-hand member.

Putting $y = x^a v$, the equation becomes

$$x^{a+1}\frac{dv}{dx} + bx^{2a}v^2 = cx^{2a},$$

or

$$\frac{dv}{c - bv^2} = \frac{dx}{x^{1-a}},$$

in which the variables are separated.

19. Summary. — In actual practice, when the equation $M\,dx + N\,dy = 0$ is to be integrated, we proceed as follows:

By inspection we can tell when

1° the variables are separated or readily separable (§ 9),

2° M and N are homogeneous and of the same degree (§§ 10, 17),

3° the equation is linear or directly reducible to one that is (§§ 13, 14),

4° M and N are linear but not homogeneous (§ 11),

5° $M = yf_1(xy)$, $N = xf_2(xy)$ (§§ 12, 17),

6° the equation is of the form $x^r y^s (my\,dx + nx\,dy) + x^\rho y^\sigma (\mu y\,dx + \nu x\,dy) = 0$ (§ 15).

If, on inspection (and with a little practice this inspection can be made very rapidly), the equation does not come under any of these heads, apply the test for an exact differential equation.† It may happen that

* This is a special form of Riccati's equation (see § 73).

† It may be possible to recognize by inspection that an equation is exact. In such case proceed at once to integrate. Or an integrating factor may be obvious by inspection (see 10° below). The general plan is to recognize, as promptly as possible, the general head under which any particular equation comes. In this summary and those of succeeding chapters the various possible methods are arranged in the order of the ease of application of the test as to whether any particular method applies.

7° $\frac{\partial M}{\partial y} - \frac{\partial N}{\partial x} = 0$ (§ 8),

8° $\frac{\partial M}{\partial y} - \frac{\partial N}{\partial x} = Nf_1(x)$ (§ 17),

9° $\frac{\partial N}{\partial x} - \frac{\partial M}{\partial y} = Mf_2(y)$ (§ 17).

If none of these cases arise, it may be possible to

10° find an integrating factor by inspection (§ 16), or to

11° find some transformation that will bring the equation under one of the above heads (§ 18).

12° As a final resort the methods of §§ 80, 25, and 72 may be tried

Ex. 1. $x\sqrt{1-y^2}\,dx + y\sqrt{1-x^2}\,dy = 0.$

Ex. 2. $\sqrt{1-y^2}\,dx + \sqrt{1-x^2}\,dy = 0.$

Ex. 3. $\frac{dy}{dx} - x^2y = x^5.$

Ex. 4. $(y-x)^2\frac{dy}{dx} = 1.$ [Put $y - x = v$].

Ex. 5. $x\frac{dy}{dx} + y + x^4y^4e^x = 0.$

Ex. 6. $(1-x)\,y\,dx + (1-y)\,x\,dy = 0.$

Ex. 7. $(y-x)\,dy + y\,dx = 0.$

Ex. 8. $x\,dy - y\,dx = \sqrt{x^2+y^2}\,dx.$

Ex. 9. $(x+a)\frac{dy}{dx} - 3y = 2(x+a)^5.$

Ex. 10. $x\,dy - y\,dx = \sqrt{x^2-y^2}\,dx.$

Ex. 11. $\left(x\sin\frac{y}{x} - y\cos\frac{y}{x}\right)dx + x\cos\frac{y}{x}\,dy = 0.$

Ex. 12. $(x-2y+5)\,dx + (2x-y+4)\,dy = 0.$

Ex. 13. $\dfrac{dy}{dx}+\dfrac{y}{(1-x^2)^{\frac{3}{2}}}=\dfrac{x+(1-x^2)^{\frac{1}{2}}}{(1-x^2)^2}.$

Ex. 14. $(1-x^2)\dfrac{dy}{dx}-xy=axy^2.$

Ex. 15. $xy^2(3y\,dx+x\,dy)-(2y\,dx-x\,dy)=0.$

Ex. 16. $(1+x^2)\dfrac{dy}{dx}+y=\tan^{-1}x.$

Ex. 17. $(5xy-3y^3)\,dx+(3x^2-7xy^2)\,dy=0.$

Ex. 18. $\dfrac{dy}{dx}+y\cos x=\dfrac{1}{2}\sin 2x.$

Ex. 19. $(xy^2+y)\,dx-x\,dy=0.$

Ex. 20. $(1-x)y\,dx-(1+y)x\,dy=0.$

Ex. 21. $3x^2y\,dx+(x^3+x^3y^2)\,dy=0.$

Ex. 22. $(x^2+y^2)(x\,dx+y\,dy)=(x^2+y^2+x)(x\,dy-y\,dx).$

Ex. 23. $(2x+3y-1)\,dx+(2x+3y-5)dy=0.$

Ex. 24. $(y^3-2x^2y)\,dx+(2xy^2-x^3)dy=0.$

Ex. 25. $(2x^3y^2-y)\,dx+(2x^2y^3-x)dy=0.$

Ex. 26. $(x^2+y^2)(x\,dx+y\,dy)+(1+x^2+y^2)^{\frac{1}{2}}(y\,dx-x\,dy)=0.$

Ex. 27. $(1+e^{\frac{x}{y}})\,dx+e^{\frac{x}{y}}\left(1-\dfrac{x}{y}\right)dy=0.$

Ex. 28. $x\,dy+(y-y^2\log x)\,dx=0.$

Ex. 29. $(x^3y^4+x^2y^3+xy^2+y)\,dx+(x^4y^3-x^3y^2-x^2y+x)\,dy=0.$

Ex. 30. $(2\sqrt{xy}-x)\,dy+y\,dx=0.$

CHAPTER III

APPLICATIONS

20. Differential Equation of a Family of Curves.— Differential equations arise in certain problems in Geometry and the physical sciences. For example, if we let x and y be the rectangular coördinates of a point in the plane, any relation among these, say $\phi(x, y) = 0$, represents some curve, and the value of $\frac{dy}{dx}$ at any point of this curve is the slope of the tangent at that point.

Starting with a relation that involves an arbitrary constant rationally

$$(1) \qquad \phi(x, y, c) = 0,$$

we have a family of curves, one curve corresponding to each value of c. The differential equation corresponding to (1)*, say

$$(2) \qquad f\left(x, y, \frac{dy}{dx}\right) = 0,$$

is of the first order. Since we can obtain from it the value of $\frac{dy}{dx}$ corresponding to any pair of values of x and y, we see that (2) defines the slope of the tangent of that curve of the family (1) which passes through any chosen point (x, y). In case (1) is of the second degree in c, we have, excluding exceptional cases, two curves of the family passing through any point, for to each pair of values of x and y correspond two values of c which are distinct, in general. If, now, we turn our attention to the differential equation (2), it must give us two values of $\frac{dy}{dx}$ for each pair of values of x and y, in general, since

* The curves defined by (1) are spoken of as the *integral curves* of (2).

two distinct curves pass through this point, and, excepting in the points where the curves are in contact, their tangents will be distinct.

If, on the other hand, we start with the differential equation and suppose that it is of the second degree in $\frac{dy}{dx}$, it is clear that since at each point of general position,* we have two values for the slope, *i.e.* two tangents to the integral curves, we must have, in general, two integral curves passing through each point. Hence it follows that the general integral involves the constant of integration to the second degree. (See footnote to Ex. 3, § 24.)

Perfectly generally, we can prove, by entirely analogous reasoning, that *the integral of a differential equation of the first order and nth degree involves the constant of integration to the nth degree.*

Ex. 1. Find the differential equation of all circles through the origin with their centres on the axis of x.

The equation of this family of circles is evidently $x^2+y^2-2ax=0$. Here a enters to the first degree.

Differentiating, we have $x+y\frac{dy}{dx}-a=0$.

Eliminating a, we get

$$2xy\frac{dy}{dx}+x^2-y^2=0, \text{ which is of the first degree.}$$

[As an exercise, the student should integrate this.]

Ex. 2. Find the differential equation of all circles of fixed radius r and with their centers on the axis of x.

The equation of this family of circles is $(x-a)^2+y^2=r^2$. Here a enters to the second degree.

Differentiating, we have, $(x-a)+y\frac{dy}{dx}=0$.

* By *point of general position*, we mean a point at which there is nothing peculiar about the family of curves (such as having two curves tangent to each other there), or about any of the curves of the family (such as a double point).

Eliminating a, we get

$$y^2\left(\frac{dy}{dx}\right)^2+y^2=r^2, \text{ which is of the second degree.}$$

Ex. 3. Find the differential equation of the system of confocal conics whose axes coincide with the axes of coördinates.

Hint.—Since the distance of the focus of the conic $\frac{x^2}{a^2}\pm\frac{y^2}{b^2}=1$ from the center is $\sqrt{a^2\mp b^2}$ it is quite clear that all the conics $\frac{x^2}{c}+\frac{y^2}{c-\lambda}=1$ have the same foci, no matter what c may be. Hence this is the equation of a system of confocal conics whose foci are at the points $(\pm\sqrt{\lambda}, 0)$. Here c is the arbitrary constant to be eliminated.

Ex. 4. Find the differential equation of the system of parabolas whose foci are at the origin of coördinates and whose axes coincide with the axis of x.

Ex. 5. Find the differential equation of the family of straight lines tangent to the circle $$x^2+y^2=r^2.$$

Ex. 6. Find the differential equation of the family of straight lines the sum of whose intercepts on the axes is a constant.

Ex. 7. Find the differential equation of the family of nodal cubics

$$(y-a)^2=2x(x-1)^2,$$

each curve of the family being tangent to the axis of y, and having its node at the point $(1, a)$.

Ex. 8. Find the differential equation of the family of nodal cubics

$$y^2=2x(x-a)^2,$$

each curve of the family being tangent to the axis of y at the origin, and having its node at the point $(a, 0)$.

Remark.—If the equation of the family of curves involves more than one arbitrary constant, the corresponding differential equation will, of course, be of higher order than the first (§ 3).

Ex. 9. Find the differential equation of all circles tangent to the axis of y.

Ex. 10. Find the differential equation of all central conics whose axes coincide with the axes of coördinates.

Ex. 11. Find the differential equation of all parabolas whose axes are parallel (a) to the axis of x, (b) to the axis of y.

Ex. 12. Find the differential equation of all circles of the same radius r.

21. Geometrical Problems involving the Solution of Differential Equations.—Differential equations of the first order arise and must be solved in geometrical problems where the curve is given by properties whose analytic expression involves the derivative of one of the coördinates of a point on the curve with respect to the other. An example or two will illustrate:

Ex. 1. Find the most general kind of curve such that the tangent at any point of it and the line joining that point with the origin (which we shall call the radius vector to that point) make an isosceles triangle with the axis of x, the latter forming the base.

The tangent of the angle between the tangent line and the axis of x is $\frac{dy}{dx}$, while that of the angle between the radius vector and the axis of x is $\frac{y}{x}$.

Hence we must have

$$\frac{dy}{dx} = -\frac{y}{x}, \text{ or } x\,dy + y\,dx = 0.$$

Integrating, we have

$$xy = \text{constant},$$

which is evidently an equilateral hyperbola.

Ex. 2. Find the most general kind of curve such that the normal at any point of it coincides, in direction, with the radius vector to that point.

Since the slope of the normal is $-\frac{dx}{dy}$, we have

$$-\frac{dx}{dy}=\frac{y}{x}, \text{ or } x\,dx+y\,dy=0.$$

Integrating, we have

$$x^2+y^2=c, \text{ a constant};$$

this is evidently a circle.

Since the differential equation is of the first degree, a single value of the constant corresponds to a pair of values of x and y. Geometrically, this means that through each point passes *one* curve of the family. Thus, if $x=1$, $y=2$, then $c=5$. That is, through the point (1, 2) passes the circle $x^2+y^2=5$, and this is the only circle of the family which does.

From the above simple examples the general method of procedure may be seen. It consists, first, in expressing analytically the given property of the curve (this gives rise to a differential equation); secondly, we must solve this equation; and finally, we must interpret geometrically the result obtained.*

In the Differential Calculus those properties of curves involving differential expressions are usually studied, and their knowledge will be presupposed here. For purpose of convenient reference, the following list will be given:

1° Rectangular coördinates

(*a*) $\frac{dy}{dx}$ is the slope of the tangent of the curve at the point (x, y);

* The general solution of the differential equation, involving an arbitrary constant, represents an infinity of curves. If we know a point through which the curve must pass, or if in any other way the conditions of the problem determine the constant of integration, either uniquely or ambiguously, one or several curves of the family alone fulfill the requirements.

(*b*) $-\dfrac{dx}{dy}$ is the slope of the normal at (x, y);

(*c*) $Y-y=\dfrac{dy}{dx}(X-x)$ is the equation of the tangent at the point (x, y), X and Y being the coördinates of any point on the line;

(*d*) $Y-y=-\dfrac{dx}{dy}(X-x)$ is the equation of the normal at (x, y);

(*e*) $x-y\dfrac{dx}{dy}$ and $y-x\dfrac{dy}{dx}$ are the intercepts of the tangent on the axes;

(*f*) $x+y\dfrac{dy}{dx}$ and $y+x\dfrac{dx}{dy}$ are the intercepts of the normal on the axes;

(*g*) $y\sqrt{1+\left(\dfrac{dx}{dy}\right)^2}$ and $x\sqrt{1+\left(\dfrac{dy}{dx}\right)^2}$ are the lengths of the tangent from the point of contact to the x and y axes respectively;

(*h*) $y\sqrt{1+\left(\dfrac{dy}{dx}\right)^2}$ and $x\sqrt{1+\left(\dfrac{dx}{dy}\right)^2}$ are the lengths of the normal from the point on the curve to the x and y axes respectively;

(*i*) $y\dfrac{dx}{dy}$ is the length of the subtangent;

(*j*) $y\dfrac{dy}{dx}$ is the length of the subnormal;

(*k*) $ds=\sqrt{dx^2+dy^2}=dx\sqrt{1+\left(\dfrac{dy}{dx}\right)^2}=dy\sqrt{1+\left(\dfrac{dx}{dy}\right)^2}$ is the element of length of arc;

(*l*) $y\,dx$ or $x\,dy$ is the element of area.

2° Polar coördinates

(*m*) $\tan\psi=\rho\dfrac{d\theta}{d\rho}$, where ψ is the angle between the radius vector and the part of the tangent to the curve drawn towards the initial line;

(*n*) $\tau=\theta+\psi$, where τ is the angle which the tangent makes with the initial line;

(*o*) $\rho \tan \psi = \rho^2 \dfrac{d\theta}{d\rho}$ is the length of the polar subtangent;

(*p*) $\rho \cot \psi = \dfrac{d\rho}{d\theta}$ is the length of the polar subnormal;

(*q*) $ds = \sqrt{d\rho^2 + \rho^2 d\theta^2} = d\rho \sqrt{1 + \rho^2\left(\dfrac{d\theta}{d\rho}\right)^2} = d\theta \sqrt{\left(\dfrac{d\rho}{d\theta}\right)^2 + \rho^2}$ is the element of length of arc;

(*r*) $\frac{1}{2} \rho^2 \, d\theta$ is the element of area;

(*s*) $p = \rho \sin \psi = \rho^2 \dfrac{d\theta}{ds}$ is the length of the perpendicular from the pole to the tangent.

(*t*) $\dfrac{1}{p^2} = \dfrac{1}{\rho^4}\left(\dfrac{d\rho}{d\theta}\right)^2 + \dfrac{1}{\rho^2}.$

Ex. 3. Determine the curves such that the normal (from the point on the curve to the axis of x) varies as the square of the ordinate. In particular find that curve which cuts the axis of y at right angles.

Using (*h*), we have the differential equation of the curve

$$y\sqrt{1 + \left(\frac{dy}{dx}\right)^2} = ky^2, \text{ or } 1 + \left(\frac{dy}{dx}\right)^2 = k^2 y^2, \text{ or } \frac{dy}{\sqrt{k^2y^2 - 1}} = dx.$$

Integrating, we get

$$\frac{1}{k}\log\left(ky + \sqrt{k^2y^2 - 1}\right) = x + c,$$

or

$$y = \frac{1}{2k}\left(ce^{kx} + \frac{1}{c}e^{-kx}\right), \text{ a family of catenaries.}$$

To find the curve of the family which cuts the axis of y at right angles, we must find that value of c for which $\dfrac{dy}{dx} = 0$ for $x = 0$. Now $\left(\dfrac{dy}{dx}\right)_{x=0} = \dfrac{1}{2}\left(c - \dfrac{1}{c}\right)$. This equals zero for $c = \pm 1$. Hence the equation of the required curve is $\pm y = \dfrac{1}{2k}(e^{kx} + e^{-kx}) = \dfrac{1}{k} \cosh kx.$

Ex. 4. Determine the curve such that the area included between an arc of it, a fixed ordinate, a variable ordinate, and the axis of x is proportional to the corresponding arc.

Using (k) and (l), we have

$$\sqrt{1+\left(\frac{dy}{dx}\right)^2}=ky, \text{ or } 1+\left(\frac{dy}{dx}\right)^2=k^2y^2.$$

This is the same differential equation that arose in Ex. 3. Hence the catenary has also this property.

Ex. 5. Find the curves such that the polar subtangent is proportional to the radius vector.

Using (o), we have $\rho^2\frac{d\theta}{d\rho}=k\rho$, or $\frac{kd\rho}{\rho}=d\theta.$

Integrating, we have $\rho^k=ce^\theta$, a family of spirals.

Ex. 6. Determine the curves whose subnormals are constant.

Ex. 7. Determine the curves whose subtangent at each point equals the square of the abscissa at that point.

Ex. 8. Determine the curves such that the perpendicular from the origin upon the tangent is equal to the abscissa of the point of contact.

Ex. 9. Determine the curves such that the angle between the radius vector and the tangent is one-half the vectorial angle.

Ex. 10. Determine the curves whose polar subtangent is four times the polar subnormal.

22. Orthogonal Trajectories. — A curve, which cuts every member of a family of curves according to some law, is called a *trajectory* of the family. Thus, it may cut every curve at a constant angle; if, in particular, that angle is a right angle, the trajectory is said to be *orthogonal.* It is at times possible to find a second family such that each curve of the one family is cut at right angles by every curve of the other. If such a pair of families of curves exists, each is said to be a *set of orthogonal trajectories* of the other.

Starting with the first family whose equation is

$$(1) \qquad \phi(x, y, c) = 0,$$

we find the corresponding differential equation

$$(2) \qquad f\left(\frac{dy}{dx}, x, y\right) = 0,$$

which, as we have noted before, defines $\frac{dy}{dx}$, the slope of the tangent at (x, y) to the curve of the family through that point. Obviously

$$(3) \qquad f\left(-\frac{dx}{dy}, x, y\right) = 0$$

will have for integral curves a family such that the slope of the tangent to the curve through the point (x, y) at this point will be the negative reciprocal of that in the case of the corresponding curve of (1); *i.e.* wherever a curve of the one family cuts one of the other their tangents are at right angles, and therefore the curves are said to be at right angles themselves.

The integral of (3) will then be the equation of the desired family.

Ex. 1. Find the orthogonal trajectories of a family of concentric circles.

Taking the common center as the origin, the equation of the circles is $x^2 + y^2 = c^2$, and their differential equation is $x + y\frac{dy}{dx} = 0$.

Hence the differential equation of their orthogonal trajectories is

$$x - y\frac{dx}{dy} = 0, \text{ or } \frac{dx}{x} - \frac{dy}{y} = 0.$$

Integrating, we have

$$y = cx,$$

a family of straight lines through the center of the circles.

Ex. 2. Find the orthogonal trajectories of the family of circles through the origin, with their centers on the axis of x.

The equation of this family of circles is $x^2+y^2-cx=0$. We have seen (Ex. 1, § 20) that the corresponding differential equation is $2xy\frac{dy}{dx}+x^2-y^2=0$. Therefore, the differential equation of the orthogonal trajectories is $2xy\frac{dx}{dy}-x^2+y^2=0$.

As this equation is obviously derivable from the other by interchanging x and y, its integral must be $x^2+y^2-cy=0$, the family of circles through the origin with their centers on the axis of y.

[Let the student verify this result by actually integrating the equation.]

Ex. 3. Show that a family of confocal central conics is self-orthogonal.

The equation of such a family of conics with their axes taken for axes of coördinates is (Ex. 3, § 20) $\frac{x^2}{c}+\frac{y^2}{c-\lambda}=1$, and its differential equation is $xy\left(\frac{dy}{dx}\right)^2+(x^2-y^2-\lambda)\frac{dy}{dx}-xy=0$.

Since this is left unaltered, when we replace $\frac{dy}{dx}$ by $-\frac{dx}{dy}$ we see that the family of curves is self-orthogonal. As a matter of fact, it is well known that a system of confocal central conics is made up of ellipses and hyperbolas, such that through any point there pass one ellipse and one hyperbola, and these cut each other at right angles. (See Ex. 17.)

Ex. 4. Prove that the family of parabolas having a common focus and a common axis is self-orthogonal.

Ex. 5. Prove that the differential equation of the family of trajectories which cut the integral curves of $f\left(\frac{dy}{dx}, x, y\right) = 0$ at an angle α is

$$f\left[\frac{\frac{dy}{dx} - \tan\alpha}{1 + \tan\alpha\,\frac{dy}{dx}}, x, y\right] = 0.*$$

In particular show that the trajectories which cut the lines $y = cx$ at a constant angle α are the logarithmic spirals

$$\frac{1}{2}\log(x^2 + y^2) + k = \frac{1}{m}\tan^{-1}\frac{y}{x},$$

or

$$r = ce^{\frac{\theta}{m}}.$$

Ex. 6. Find the trajectories which cut at a constant angle α the circles through the origin with their centers on the axis of x.

Ex. 7. Find the trajectories which cut at a constant angle α (other than a right angle) a system of concentric circles.

Ex. 8. Find the orthogonal trajectories of the parabolas $y^2 = 4cx$.

Ex. 9. Find the orthogonal trajectories of the hyperbolas $x^2 - y^2 = c$.

Ex. 10. Find the orthogonal trajectories of the similar central conics $ax^2 + by^2 = c$, where a and b are fixed constants and c the arbitrary constant.

In polar coördinates the equation of a family of curves will be

$$(1') \qquad \phi(\rho, \theta, c) = 0;$$

and the corresponding differential equation will be

$$(2') \qquad f\left(\frac{d\theta}{d\rho}, \rho, \theta\right) = 0,$$

which defines $\frac{d\theta}{d\rho}$ at each point (ρ, θ).

* In practice it will be simpler to replace $\tan\alpha$ by a single letter, say m.

We have noted, § 21 (m), that, if ψ is the angle between the radius vector to the point (ρ, θ) and the tangent at that point drawn in a definite direction, $\tan\psi = \rho\dfrac{d\theta}{d\rho}$. If now, ψ' is the angle for the curve cutting this one at right angles at the point (ρ, θ), we have $\psi' - \psi = \pm\dfrac{\pi}{2}$, or $\psi' = \psi \pm \dfrac{\pi}{2}$. Hence

$$\tan\psi' = -\cot\psi = -\frac{1}{\tan\psi}.$$

Using primed letters for the second curve, we have then

$$\rho'\frac{d\theta'}{d\rho'} = -\frac{1}{\rho}\frac{d\rho}{d\theta}, \text{ whence } \frac{d\theta}{d\rho} = \frac{-1}{\rho\rho'}\frac{d\rho'}{d\theta'}.$$

At the point of intersection of the two curves $\rho' = \rho$, $\theta' = \theta$. Hence

(3′) $$f\left(\frac{-1}{\rho^2}\frac{d\rho}{d\theta}, \rho, \theta\right) = 0$$

is the differential equation of the orthogonal trajectories of (1′), since the value of $\rho\dfrac{d\theta}{d\rho}$ given by it equals that of $-\dfrac{1}{\rho}\dfrac{d\rho}{d\theta}$ given by (2′).

Note.— Frequently θ is taken as the independent variable. In this case (2′) will be in the form $f\left(\dfrac{d\rho}{d\theta}, \rho, \theta\right) = 0$, and (3′) will then be

$$f\left(-\rho^2\frac{d\theta}{d\rho}, \rho, \theta\right) = 0.$$

Ex. **11.** Find the orthogonal trajectories of the family of lemniscates $\rho^2 = c\cos 2\theta$.

Differentiating and eliminating c, we find the differential equation of the family to be $\dfrac{1}{\rho}\dfrac{d\rho}{d\theta} = -\tan 2\theta$. Hence the differential equation of the orthogonal trajectories is

$$-\rho\frac{d\theta}{d\rho} = -\tan 2\,\theta,$$

or

$$\frac{d\rho}{\rho} = \cot 2\,\theta\, d\theta.$$

Integrating, we find $\rho^2 = k \sin 2\,\theta$, a second family of lemniscates whose axis makes an angle of $45°$ with that of the first family.

Ex. **12.** Find the orthogonal trajectories of the family of cardioids $\rho = c(1 - \cos\theta)$.

Ex. **13.** Find the orthogonal trajectories of the family of logarithmic spirals $\rho = e^{c\theta}$.

Ex. **14.** Find the orthogonal trajectories of the family of curves $\rho^m \sin m\theta = c^m$.

Ex. **15.** Find the orthogonal trajectories of the family of curves $\frac{1}{\rho} = \sin^2\theta + c$.

Ex. **16.** Find the orthogonal trajectories of the family of confocal and coaxial parabolas $\rho = \frac{2\,c}{1 - \cos\theta}$.

Ex. **17.** Find the orthogonal trajectories of the family of confocal conics $\rho = \frac{c^2 - \lambda^2}{c - \lambda\cos\theta}$, c being the parameter, λ a fixed constant.

23. Physical Problems giving Rise to Differential Equations. — Problems frequently arise in Mechanics, Electricity, and other branches of Physics whose solution involves the solution of differential equations. As a knowledge of these subjects is necessary to understand properly the problems that arise in them, we shall restrict ourselves, as far as possible, to problems involving only very elementary principles. As in the case of geometrical problems, the mode of procedure is, first the analytic expression of the given data of the problem, (this gives rise to the differential equation); then comes the problem of solving this equation, with an interpretation of the result. Frequently there is the additional step of fixing the value of the constant of integration so as to satisfy the requirements of the problem.

The following examples will illustrate:

Ex. 1. A body falls vertically, acted upon by gravity only. If it has an initial velocity v_0, what will be its velocity at any given instant, and what will be the distance covered in any given period of time?

The motion being rectilinear, a single coördinate, x, will be sufficient to determine the position of the body. Suppose the position of the body at the time $t = 0$ to be x_0 (this is called its initial position).

The velocity, which we shall represent by v, is $\frac{dx}{dt}$, and the acceleration, represented by f, is $\frac{dv}{dt}$. In case gravity acts alone, the acceleration is constant, and this constant is usually designated by g. We have now

$$\frac{dv}{dt} = g.$$

Integrating this equation we get

$$v = gt + c.$$

When $t = 0$, we have given $v = v_0$. $\therefore c = v_0$, and the answer to our first question is given by

$$v = gt + v_0.$$

To find the position of the body at any instant we must integrate

$$\frac{dx}{dt} = gt + v_0.$$

The general solution of this is

$$x = \frac{1}{2} gt^2 + v_0 t + c.$$

Since $x = x_0$ when $t = 0$, we must have $c = x_0$. Hence at any instant t,

$$x = \tfrac{1}{2} gt^2 + v_0 t + x_0,$$

and the distance covered in the period t is

$$x - x_0 = \tfrac{1}{2} gt^2 + v_0 t.$$

Ex. 2. A particle descends a smooth plane making an angle α with the horizontal plane. The only force acting is gravity. If the particle starts from rest, find the velocity at any moment t, and the distance traveled in the time t. [*Hint.* — Here the acceleration is $g \sin \alpha$, the component of g in the direction of the motion.] Prove that if a particle starts at rest from the highest point of a vertical circle, it will reach any other point of the circle when moving along the chord to that point in the same time it would take to drop to the lowest point of the circle.—TAIT AND STEELE, *Dynamics of a Particle.*

Ex. 3. A particle falls through a resisting medium (such as air) in which the resistance is proportional to the square of the velocity; what is its motion?

The equation of motion is then

$$\frac{dv}{dt} = g - kv^2.$$

Here the variables are separable, and we have

$$\frac{dv}{g - kv^2} = dt.$$

Putting $gk = r^2$, this becomes

$$\frac{g\,dv}{g^2 - r^2v^2} = dt, \text{ or } \frac{dv}{g + rv} + \frac{dv}{g - rv} = 2\,dt.$$

Integrating $\displaystyle\int_{v_0}^{v} \left(\frac{dv}{g + rv} + \frac{dv}{g - rv} \right) = 2 \int_0^t dt$, we have*

* A knowledge of hyperbolic functions enables one to effect the integration much more expeditiously, especially if $v_0 = 0$.

$$\int_0^v \frac{g\,dv}{g^2 - r^2v^2} = \frac{1}{r} \tanh^{-1} \frac{rv}{g}. \quad \therefore \; v = \frac{dx}{dt} = \frac{g}{r} \tanh rt.$$

Integrating again, we have $x - x_0 = \dfrac{g}{r^2} \log \cosh rt.$

$$\frac{1}{r}[\log(g+rv)-\log(g+rv_0)-\log(g-rv)+\log(g-rv_0)]=2t,$$

or
$$\frac{1}{r}\log\left(\frac{g-rv_0}{g+rv_0}\,\frac{g+rv}{g-rv}\right)=2t.$$

$$\therefore \frac{g+rv}{g-rv}=\frac{g+rv_0}{g-rv_0}e^{2rt}.$$

Temporarily put the constant $\frac{g+rv_0}{g-rv_0}=c.$ Then

$$v=\frac{g}{r}\cdot\frac{ce^{2rt}-1}{ce^{2rt}+1}.$$

Since $v=\frac{dx}{dt}$, the position at any time is given by

$\int_{x_0}^{x}dx=x-x_0=\frac{g}{r}\int_0^t\frac{ce^{2rt}-1}{ce^{2rt}+1}dt$, where c must finally be replaced by its value in terms of v_0. If the body falls from rest, $v_0=0$, $\therefore c=1$.

[As an exercise, the student may carry out the integration indicated above.]

Ex. 4. The acceleration of a particle moving in a straight line is proportional to the cube of the velocity and in the opposite direction from the latter. Find the distance passed over in the time t, the initial velocity being v_0, and the distance being measured from the initial position of the particle, *i.e.* $x_0=0$.

$$\frac{dv}{dt}=-kv^3, \text{ or } \frac{dv}{-v^3}=k\,dt.$$

$$\therefore \frac{1}{v^2}-\frac{1}{v_0^2}=2kt, \text{ or } v=\frac{v_0}{\sqrt{2kv_0^2t+1}}=\frac{dx}{dt};$$

and
$$x=\frac{\sqrt{2kv_0^2t+1}-1}{kv_0}.$$

Ex. 5. Find the distance passed over in the time t, if the acceleration is proportional to the velocity.

Ex. **6.** One of the important equations in the theory of electricity is

$$L\frac{di}{dt} + Ri = E,$$

where i is the current, L the coefficient of self-induction (a constant), R the resistance (a constant), and E the electromotive force, which may be a constant (including zero) or a function of the time. This equation is linear (§ 13). Find i if (*a*) $E = 0$, (*b*) $E =$ constant, (*c*) $E = E_0 \sin \omega t$, (E_0 and ω being constants), (*d*) $E =$ any function of the time, say $E(t)$.

Case (*c*) plays such an important rôle in the Theory of Electricity that its solution is given here in detail. This equation arises in the case of alternating currents, where the electromotive force is a periodic function of the time, the period being $\frac{2\pi}{\omega}$, and the maximum value of the electromotive force is E_0.

$$\frac{di}{dt} + \frac{R}{L} i = \frac{E_0}{L} \sin \omega t.$$

An integrating factor is $e^{\frac{R}{L}t}$.

$$\therefore\ ie^{\frac{R}{L}t} = \frac{E_0}{L}\int e^{\frac{R}{L}t} \sin \omega t\, dt + c.$$

Since $\int e^{at} \sin \omega t\, dt = \dfrac{a \sin \omega t - \omega \cos \omega t}{a^2 + \omega^2} e^{at}$,

$$ie^{\frac{R}{L}t} = \frac{E_0}{L} \frac{\frac{R}{L}\sin \omega t - \omega \cos \omega t}{\frac{R^2}{L^2} + \omega^2} e^{\frac{R}{L}t} + c$$

$$= E_0\left(\frac{R \sin \omega t - \omega L \cos \omega t}{R^2 + \omega^2 L^2}\right) e^{\frac{R}{L}t} + c$$

$$=\frac{E_0}{\sqrt{R^2+\omega^2L^2}}\,\frac{R\sin\omega t-\omega L\cos\omega t}{\sqrt{R^2+\omega^2L^2}}e^{\frac{R}{L}t}+c$$

$$=\frac{E_0\,e^{\frac{R}{L}t}}{\sqrt{R^2+\omega^2L^2}}\sin(\omega t-\phi)+c,$$

where
$$\sin\phi=\frac{\omega L}{\sqrt{R^2+\omega^2L^2}},\quad \cos\phi=\frac{R}{\sqrt{R^2+\omega^2L^2}}.$$

$$\therefore i=\frac{E_0}{\sqrt{R^2+\omega^2L^2}}\sin(\omega t-\phi)+ce^{-\frac{R}{L}t}.$$

The term $ce^{-\frac{R}{L}t}$ usually becomes negligible after a very short interval of time. The current then becomes periodic with the same frequency as the electromotive force. But the two are not in the same phase, the current lagging behind by the angle ϕ.

CHAPTER IV

DIFFERENTIAL EQUATIONS OF THE FIRST ORDER AND HIGHER DEGREE THAN THE FIRST

24. Equations Solvable for p. — For the sake of simplicity we shall adopt the generally accepted notation of replacing $\frac{dy}{dx}$ by p. Our equation may then be written

$$(1) \qquad f(p, x, y) = 0.$$

If this equation is of the nth degree, we may look upon it as an algebraic equation of the nth degree in p. Let its roots be $f_1(x, y)$, $f_2(x, y), \cdots, f_n(x, y)$, then the equation may also be written

$$(2) \qquad [p - f_1(x, y)][p - f_2(x, y)] \cdots [p - f_n(x, y)] = 0.$$

Consider now the differential equations arising on equating each of these factors separately to zero. If we can integrate each of these by some method of Chapter II, we can readily get the general solution of (1). Let the solutions of the separate equations arising from (2) be $\phi_1(x, y, c) = 0$, $\phi_2(x, y, c) = 0, \cdots, \phi_n(x, y, c) = 0$. It is quite clear that

$$(3) \qquad \phi_1(x, y, c)\,\phi_2(x, y, c)\,\phi_3(x, y, c) \cdots \phi_n(x, y, c) = 0^*$$

is the solution of (1). For, the vanishing of the left-hand member of (3) means the vanishing of one of its factors. This will cause one of the factors of (2) to vanish, when substituted in it, and consequently, the original equation will be satisfied. Besides, a constant of inte-

* If one of the ϕ's is not rational, there will be found certain other irrational ones which, with it, form a set of conjugate irrational functions whose product is rational (see Ex. 2 below). The result of rationalizing any ϕ of the set is the same as this rational product. In practice it is frequently desirable to make use of this fact. The student should verify this fact in the case of Ex. 2.

gration is involved.* We see that the constant enters to the same degree in (3) that p does in (1). (Theorem, § 20.)

Ex. 1. $p^2 + (x+y)p + xy = 0$,

or $$(p+x)(p+y) = 0.$$

Integrating $\frac{dy}{dx} + x = 0$, we get $2y + x^2 = c$;

and integrating $\frac{dy}{dx} + y = 0$, we get $\log y + x = k$, or $y = ce^{-x}$.

Hence the general solution is $(2y + x^2 - c)(y - ce^{-x}) = 0$.

Ex. 2. $xp^2 - 2yp - x = 0$.

Solving this for p, we get $p = \frac{y \pm \sqrt{x^2+y^2}}{x}$. We have, then, to consider the two equations $xp - y = \sqrt{x^2+y^2}$

and $$xp - y = -\sqrt{x^2+y^2}.$$

$\frac{x\,dy - y\,dx}{\sqrt{x^2+y^2}} = dx$ may be written $\frac{x\,dy - y\,dx}{x^2\sqrt{1+\left(\frac{y}{x}\right)^2}} = \frac{dx}{x}$, which gives,

on integrating,

$$\log\left[\frac{y}{x} + \sqrt{1+\left(\frac{y}{x}\right)^2}\right] = \log x + k, \text{ or } \frac{y}{x} + \sqrt{1+\left(\frac{y}{x}\right)^2} - cx = 0.$$

* Since (3) is a solution by having each of its factors separately satisfying the differential equation (1), it may be asked why we use the same constant c in all the factors. If we look upon equation (1) as equivalent to the n separate equations arising on equating each of the factors of (2) to zero (which, in fact, it is), then we really have n equations to solve, and the various factors of (3), involving distinct constants, are the solutions of these. But if we require the general solution of (1) to be given as a single expression, first, there is no room for more than one constant of integration (§ 4), and then, there is no loss in using the single constant throughout, from the very fact that (3) is a solution by virtue of each factor separately satisfying the equation.

Integrating $xp - y = -\sqrt{x^2+y^2}$, we get in a like manner

$$-\log\left[\frac{y}{x}+\sqrt{1+\left(\frac{y}{x}\right)^2}\right]=\log\left\{\frac{1}{\frac{y}{x}+\sqrt{1+\left(\frac{y}{x}\right)^2}}\right\}$$

$$=\log\left\{\frac{\frac{y}{x}-\sqrt{1+\left(\frac{y}{x}\right)^2}}{-1}\right\}=\log x+k, \text{ or } \frac{y}{x}-\sqrt{1+\left(\frac{y}{x}\right)^2}-cx=0.$$

Hence the solution is

$$\left[\frac{y}{x}+\sqrt{1+\left(\frac{y}{x}\right)^2}-cx\right]\left[\frac{y}{x}-\sqrt{1+\left(\frac{y}{x}\right)^2}-cx\right]=0,$$

or
$$c^2x^2-2cy-1=0.$$

Ex. 3. $y^2+p^2=1.$

Here $p=\pm\sqrt{1-y^2}$. Solution of $\frac{dy}{\sqrt{1-y^2}}=dx$ is $\sin^{-1}y=x+c$, or $y=\sin(x+c)$; while that of $\frac{dy}{-\sqrt{1-y^2}}=dx$ is $y=\cos(x+c)$. Since c is perfectly arbitrary, either one of these is sufficient as the general solution of the equation; * for $\sin\left(x+c+\frac{\pi}{2}\right)=\cos(x+c)$.

* Since c enters in a transcendental way, and not to the second degree in the solution $y=\sin(x+c)$, or $y=\cos(x+c)$, we seem to have an exception to the rule of § 20. It is really not such. That rule presupposes that the constant of integration enters algebraically. We can make our solution conform to the rule by writing it in the form $(\sin^{-1}y-x-c)(\cos^{-1}y-x-c)=0$. As a matter of fact the rule was based on the fact that through each point of general position pass two integral curves of a differential equation of the second degree. So here, of the infinite number of values of c that satisfy $y=\sin(x+c)$ for a given pair of values of x and y, only two will determine distinct curves. A more elegant form than the one above, in which c enters algebraically and to the second degree, may be gotten thus: Since c is any number, we may write $\sin c=\frac{1-k^2}{1+k^2}$; then $\cos c=\frac{2k}{1+k^2}$; whence, remembering that $\sin(x+c)=\cos c\sin x+\sin c\cos x$, the solution takes the form

$$y-\cos x-2k\sin x+k^2(y+\cos x)=0.$$

Ex. 4. $(2xp - y)^2 = 8x^3$.

Ex. 5. $(1 + x^2)p^2 = 1$.

Ex. 6. $p^3 - (2x + y^2)p^2 + (x^2 - y^2 + 2xy^2)p - (x^2 - y^2)y^2 = 0$.

25. Equations Solvable for y. — If the equation can be solved for y, the following method will frequently be found useful. Solving for y, we have

$$y = \psi(x, p). \tag{4}$$

Differentiating this we get

$$p = \frac{\partial \psi}{\partial x} + \frac{\partial \psi}{\partial p}\frac{dp}{dx}, \tag{5}$$

a differential equation which is really of the second order, but since y is no longer present, it may be looked upon as an equation of the first order in the variables x and p. It may happen that we can integrate this equation. Suppose its solution to be

$$\omega(x, p) = c. \tag{6}$$

Eliminating p from (4) and (6) we have

$$\phi(x, y, c) = 0, \tag{7}$$

which is a solution of (4); and, since it involves an arbitrary constant, it is the general solution.

We know that (7) is the solution of (4) from the following considerations: Since (5) is the derivative of (4), every solution of (4) is a solution of (5), considered as an equation of the second order in x and y. Since (6) is a solution of (5), every solution of (6) looked upon as a differential equation in x and y is a solution of (5). (4) and (6) are known as *first integrals* of (5). Since (4) contains y and (6) does not, these two first integrals are evidently independent. Equation (4) has an infinite number of integral curves,* and so has

* We use this geometrical mode of expression, not because it is essential to the argument, but because it is simpler.

(6) for each value of c. To find the curve or curves common to the two families of integral curves we shall have to find the equation of the locus of those points for which (4) and (6) determine the same value of p.* But this locus is evidently gotten by eliminating p from (4) and (6). For each value of c we get thus one integral curve of (4). Hence, when c is an arbitrary constant, we have the general solution.†

Note. — This method applies equally well to equations of the first degree. See Ex. 4.

Remark. — At times it is easy to integrate (6). Doing this, a relation

(8) $$\Phi\,(x, y, c, c') = 0,$$

involving two constants, results. This is the general solution of (5) and must therefore contain that of (4) for some relation between c and c'. This relation can be found by substituting (8) in (4) and noting what condition is imposed, so that the equation be satisfied. In actual practice, however, this method will generally not be as desirable as the one given above.

Ex. 1. $2px - y + \log p = 0$; or

(1) $$y = 2px + \log p.$$

Differentiating, we get

$$p = 2p + \left(2x + \frac{1}{p}\right)\frac{dp}{dx},$$

or

$$p\,dx + 2x\,dp + \frac{1}{p}\,dp = 0.$$

An integrating factor is seen, by inspection, to be p. Using this, we have

$$p^2\,dx + 2px\,dp + dp = 0.$$

Integrating we have

(2) $$p^2x + p = c.$$

Eliminating p between (1) and (2), we have the required result.

* As already mentioned (§ 20), a differential equation of the first order may be looked upon as defining p, the slope of the integral curve, at each point (x, y).

† Since the process of eliminating p from (4) and (6) may introduce extraneous factors, and errors may enter in other ways, it is desirable to test the result (7), by finding out whether it actually satisfies the equation (4).

Remark. — In this case, while it is perfectly possible to perform this elimination [since (2) can be readily solved for p, and the latter value can be put in (1)], the result will not be very attractive in form. It is simpler to say that (1) and (2) taken together constitute the solution, in that, from them, we can express x and y in terms of p, which may be looked upon as a parameter. Thus from (2) we have

$$\left.\begin{aligned} x &= \frac{c-p}{p^2}, \\ \text{and } \therefore \quad y &= \frac{2(c-p)}{p} + \log p. \end{aligned}\right\}$$

Such parametric representation is frequently resorted to; thus, for example, the parametric equations of the ellipse

$$\left\{\begin{aligned} x &= a \cos \theta, \\ y &= b \sin \theta, \end{aligned}\right\}$$

where θ is the eccentric angle, and also the usually adopted equations of the cycloid

$$\left\{\begin{aligned} x &= a(\theta - \sin \theta), \\ y &= a(1 - \cos \theta). \end{aligned}\right\}$$

Ex. 2. $4xp^2 + 2xp - y = 0$;

or $$y = 2xp + 4xp^2.$$

Differentiating, we have on collecting terms,

$$(4p+1)\left(2x\frac{dp}{dx} + p\right) = 0.$$

Neglecting the factor $4p+1$, whose significance we shall see later (§ 32, note), we have, integrating the other factor, $xp^2 = k^2$.

Hence the solution is $y = 2k\sqrt{x} + 4k^2$, or putting $2k = c$,

$$y = c\sqrt{x} + c^2.$$

Rationalizing this, we have

$$(y - c^2)^2 = c^2x, \text{ or putting } c^2 = C,$$

$$(y - C)^2 = Cx.$$

Ex. 3. $xp^2 - 2yp - x = 0$. (Ex. 2, § 24.)

Ex. 4. $p + 2xy = x^2 + y^2$.

Ex. 5. $y = -xp + x^4p^2$.

Ex. 6. $p^2 + 2xp - y = 0$.

26. Equations Solvable for x.—A method, entirely analogous to that of the previous paragraph, can be deduced in case the equation can be solved for x. Suppose the equation in the form

$$x = \theta(y, p). \tag{9}$$

Differentiating with respect to y we get

$$\frac{dx}{dy} = \frac{1}{p} = \frac{\partial\theta}{\partial y} + \frac{\partial\theta}{\partial p}\frac{dp}{dy}. \tag{10}$$

Here x no longer appears, and we may look upon this equation as one of the first order in y and p. If we can integrate this, we obtain a relation involving an arbitrary constant,

$$\chi(y, p, c) = 0; \tag{11}$$

and on eliminating p between (9) and (11) we have the general solution.

Ex. 1. $x + py(2p^2 + 3) = 0$.

Ex. 2. $a^2yp^2 - 2xp + y = 0$.

Ex. 3. $xp^2 - 2yp - x = 0$. (Ex. 3, § 25.)

Ex. 4. $p^3 - 4xyp + 8y^2 = 0$.

Ex. 5. Find the family of curves for which the length of the normal (from the curve to the axis of x) is equal to the square root of the length of its intercept on the axis of x.

27. Clairaut's Equation.* — If the equation is of the form

$$(1) \qquad y = px + f(p),$$

where $f(p)$ is any function of p, the solution is gotten so readily that especial attention should be given to this form of the equation, in order that it may be recognized at once.†

Using the method of § 25, we have

$$p = p + [x + f'(p)]\frac{dp}{dx},$$

or

$$\frac{dp}{dx}[x + f'(p)] = 0.$$

Neglecting the factor $x + f'(p)$, which involves no differential expressions (see § 32, note), we have

$$\frac{dp}{dx} = 0, \text{ whence } p = c.$$

Putting this value in (1) we have

$$(2) \qquad y = cx + f(c),$$

which is the general solution of (1).

Ex. 1. $(px - y)^2 = p^2 + 1$.

Solving for y, $\qquad y = px \pm \sqrt{p^2 + 1}$.

This being in Clairaut's form, its solution is known at once to be

$$y = cx \pm \sqrt{c^2 + 1},$$

or

$$(cx - y)^2 = c^2 + 1.$$

* This form of equation is named after Alexis Claude Clairaut (1713–1765). He was the first to apply the process of differentiation (§§ 25, 26) to the solution of equations. His application of this method to the equation that bears his name was published, *Histoire de l'Académie des Sciences de Paris*, 1734.

† The student should be able to recognize this equation, not only when it is solved for y, as it is in (1). Obviously, what characterizes this form of the equation is that x and y occur only in the combination $y - px$. Hence any function of $y - px$ and p equated to zero, say $f(y - px, p) = 0$ is a Clairaut equation, and its solution is $f(y - cx, c) = 0$. (See Ex. 1.)

Ex. 2. $4e^{2y}p^2 + 2xp - 1 = 0$.

Put $e^{2y} = t$, then $p = \frac{dy}{dx} = \frac{1}{2t}\frac{dt}{dx}$, and the equation becomes

$$t = x\frac{dt}{dx} + \left(\frac{dt}{dx}\right)^2.$$

Its solution is $t = cx + c^2$, or $e^{2y} = cx + c^2$.

Remark.—At times, as in the case above, a transformation can be found to simplify very materially an equation which will not yield directly to any of the previous methods. Unfortunately these transformations are not always obvious. Experience, and frequently that alone, will help one in making a proper choice.

Ex. 3. $4e^{2y}p^2 + 2e^{2x}p - e^{2x} = 0$.

Put $e^x = u$, $e^{2y} = v$, then $p = \frac{u}{2v}\frac{dv}{du}$, and the equation becomes

$$v = u\frac{dv}{du} + \left(\frac{dv}{du}\right)^2,$$

whence its solution is $e^{2y} = ce^x + c^2$.

Note.—One's first impulse would be to try $e^{2x} = u$ and $e^{2y} = v$. Our equation then becomes $v = 2u\frac{dv}{du} + 4u\left(\frac{dv}{du}\right)^2$. This is not in Clairaut's form; but it can be integrated (see Ex. 2, § 25), so that this transformation is also effective.

Ex. 4. $e^{2y}p^3 + (e^{2x} + e^{3x})p - e^{3x} = 0$.

Ex. 5. $xy^2p^2 - y^3p + x = 0$. (Let $x^2 = u$, $y^2 = v$.)

Ex. 6. $(x^2 + y^2)(1 + p)^2 - 2(x + y)(1 + p)(x + yp) + (x + yp)^2 = 0$. (Let $x + y = u$, $x^2 + y^2 = v$.)

Ex. 7. $y = 2px + y^2p^3$. (Let $y^2 = v$.)

Ex. 8. $a^2yp^2 - 2xp + y = 0$. (Ex. 2, § 26.) (Let $2x = u$, $y^2 = v$.)

Ex. 9. $(xp - y)^2 = x^2(2xy - x^2p)$.

28. Summary. — Given a differential equation of the first order and higher degree than the first, there are three methods * which suggest themselves, to be tried in actual practice in the following order:

1° Solve for p, and then solve the resulting equations of the first degree (§ 24).

2° Solve for y, differentiate with respect to x, integrate, and eliminate p between this solution and original equation (§ 25).

3° Solve for x, differentiate with respect to y, integrate, and eliminate p between this solution and original equation (§ 26).

Clairaut's form (§ 27) has been given special prominence in this chapter because of the ease of finding its solution. It is, of course, solved by method 2°.

If none of the above methods work, a substitution must be sought to bring the equation into manageable shape.

There are certain cases when we can tell in advance that some or all of these methods work. Here the difficulties are those of Algebra or the Integral Calculus, and not of the Differential Equations. For example, consider the following cases:

(*a*) If the equation in p is algebraic and all the coefficients are homogeneous and of the same degree in x and y, then, on dividing by the leading coefficient, all the coefficients are homogeneous and of zero degree. Hence, if we can solve for p (which is an algebraic process), we shall find p as homogeneous functions of x and y of degree zero, and the resulting equations, when subjected to the transformation $y = vx$, will have their variables separated (§ 10) and are solvable by quadratures.

Again, since after dividing by one of the coefficients of the equation the equation is a function of p and $\frac{y}{x}$, say $f\left(p, \frac{y}{x}\right) = 0$, if we solve for y, $\left(\text{or } \frac{y}{x}\right)$, we get $y = x\psi(p)$. Differentiating, we have,

$$p = \psi(p) + x\psi'(p)\frac{dp}{dx}, \text{ or } \frac{\psi' dp}{p - \psi(p)} = \frac{dx}{x},$$

where the variables are separated.

* It is almost needless to remark that these methods are not mutually exclusive. Two, or all three, methods may be applicable to some equations.

Hence, in this case methods 1° and 2° both work, provided we can solve for p and for y.

(b) If x is absent, so that the equation is of the form $f(p, y) = 0$, solving for p, we get $p = \frac{dy}{dx} = \psi(y)$, or $\frac{dy}{\psi(y)} = dx$.

Again, solving for y, we get $y = \psi(p)$, and differentiating, we have $p = \psi'(p)\frac{dp}{dx}$, or $\frac{\psi'(p)\,dp}{p} = dx$, where the variables are separated.

Here again methods 1° and 2° both work.

(c) If y is absent, equation is $f(p, x) = 0$. Let the student show, as an exercise, that in this case methods 1° and 3° both work, provided we can solve for p and for x.

(d) If the equation is of the first degree in x and y, thus, $xf_1(p) + yf_2(p) + f_3(p) = 0$*, it is readily seen that method 2° works. For, solving for y, we have

$$y = x\psi_1(p) + \psi_2(p).$$

Differentiating, we get

$$p = \psi_1(p) + [x\psi_1'(p) + \psi_2'(p)]\frac{dp}{dx}.$$

Considering p as the independent variable, this may be written,

$$\frac{dx}{dp} + x\frac{\psi_1'(p)}{\psi_1(p) - p} = \frac{\psi_2'(p)}{p - \psi_1(p)},$$

which is linear and can be solved by quadratures (§ 13).

Ex. 1. $y^2(1 + p^2) = a^2$.

Ex. 2. $yp = (x - b)p^2 + a$.

Ex. 3. $x^3p^2 + x^2yp + 1 = 0$.

Ex. 4. $3p^2x - 6yp + x + 2y = 0$.

Ex. 5. $y = p^2(x + 1)$.

Ex. 6. $(px - y)(py + x) = a^2p$. (Let $x^2 = u$, $y^2 = v$.)

Ex. 7. $p^2 + 2py\cot x = y^2$.

* Clairaut's equation is a special case of this.

Ex. 8. $(1+x^2)p^2 - 2xyp + y^2 - 1 = 0.$

Ex. 9. $x^2p^2 - 2(xy + 2p)p + y^2 = 0.$

Ex. 10. $y = xp + \dfrac{yp^2}{x^2}.$ (Let $x^2 = u$, $y^2 = v$.)

Ex. 11. $x^2p^2 - 2xyp + y^2 = x^2y^2 + x^4.$

Ex. 12. $\dfrac{y - xp}{\sqrt{1+p^2}} = f(\sqrt{x^2+y^2}).$ (Let $x = \rho\cos\theta$, $y = \rho\sin\theta$.)

Ex. 13. Find the equation of the curves for which the distance of the tangent from the origin varies as the distance of the point of contact from the origin.

Ex. 14. Find the equation of the curves such that the square of the length of arc measured from a fixed point is a constant times the ordinate of the point. (Let the constant factor be $4k$.)

Ex. 15. Find the equation of the curves down each of whose tangents a particle, starting from rest, will slide to the horizontal axis in the same time. [As in Ex. 2, § 23, we have that the distance covered in the time t equals $\frac{1}{2}gt^2\sin\alpha$. Here $\sin\alpha = \dfrac{p}{\sqrt{1+p^2}}$, and the distance covered is $\dfrac{y\sqrt{1+p^2}}{p}$, § 21, (g).]

CHAPTER V

SINGULAR SOLUTIONS

29. Envelopes. — We have noticed before that $\phi(x, y, c) = 0$, where c is an arbitrary constant, represents a family of curves, to each value of c corresponding some definite curve (provided c enters rationally, which we shall suppose to be the case throughout this chapter). So that if we pick out some curve corresponding to a definite value of c, we can suppose our attention directed to the different curves corresponding to c as it varies continuously.*

We shall be interested in the locus of the ultimate points of intersection of each curve with its consecutive one. By the *ultimate points of intersection of a curve with its consecutive one* we mean the limiting positions of the points of intersection of a curve with a neighboring one as the latter approaches coincidence with the former. (Thus in the case of the family of circles referred to in the footnote, the ultimate points of intersection of two consecutive curves are the extremities of the diameter perpendicular to the axis of x.) To find the equation of this locus, we proceed as follows: If

$$(1) \qquad \phi(x, y, c) = 0$$

is the equation of a curve corresponding to some chosen value of c,

$$(2) \qquad \phi(x, y, c + \Delta c) = 0$$

* Thus, for example, consider the family of circles of fixed radius r whose centers all lie on the axis of x; their equation is $(x - c)^2 + y^2 = r^2$. When $c = 0$ we have the circle whose center is at the origin, and as c increases we get circle after circle whose center is $(c, 0)$.

will be the equation of a neighboring curve, Δc being a finite constant quantity, different from zero. To find the points of intersection of these two curves we have to solve (1) and (2) for x and y, or we may replace (2) by a constant times their difference, *i.e.* by

$$(3) \qquad \frac{\phi(x, y, c+\Delta c)-\phi(x, y, c)}{\Delta c}=0.$$

To obtain the ultimate points of intersection of the curve with its consecutive one we combine (1) with what (3) becomes when we let Δc approach the limit 0, *i.e.* with

$$(4) \qquad \frac{\partial \phi(x, y, c)}{\partial c}=0.$$

If we were to solve (1) and (4), we should actually obtain the intersections of the curve with the consecutive one. But what we want

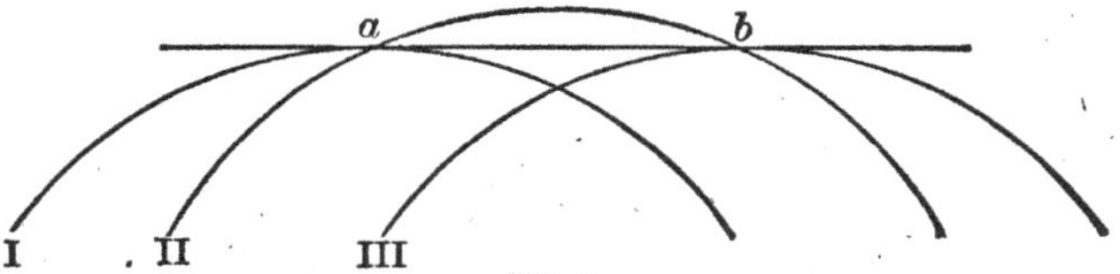

FIG. 1

is the locus of these points, for all values of c. This is evidently gotten by eliminating c between (1) and (4). This locus is known as the *envelope* of the family (1). A property of the envelope which we shall have occasion to use is: *At each point of the envelope there is one curve of the family tangent to it.* This is immediately obvious from the figure. Suppose (I), (II), (III) are three curves of the family which ultimately become coincident. a becomes an ultimate point of intersection of (I) and (II), and b of (II) and (III). Hence they are both points on the envelope, and the line joining them becomes ultimately a tangent to the envelope. But they are also both on the curve (II), so that the line joining them also becomes a tangent to (II). That is to say, (II) is tangent to the envelope at the limiting position of a and b.*

Ex. Find the envelope of the family of circles referred to in the footnote, page 61.

30. Singular Solutions. — Suppose now that

(1) $$\phi(x, y, c) = 0$$

is the solution of $f(p, x, y) = 0$. We have already noted that, looked at geometrically, this means that the slope of the tangent at a point (x, y) of the curve of the family defined by (1) passing through that point is exactly the value of p given by $f(p, x, y) = 0$ for that pair of values of x and y. But we have just seen that the tangent at any point of the envelope of the family of integral curves coincides with that of the integral curve through that point. It follows, then, that the equation of the envelope will satisfy the differential equation, and is consequently a solution. Moreover, since the envelope is usually not a curve of this family, *i.e.* its equation cannot be gotten from (1) by assigning a definite value to the parameter, the equation of this envelope is a solution, distinct from the general solution. It contains no arbitrary constant, and is not a particular solution. It is known as the *singular solution*.

Ex. 1. $y = px + \frac{1}{p}$.

This being Clairaut's equation, its solution is

$$y = cx + \frac{1}{c},$$

or

$$c^2x - cy + 1 = 0.$$

* This theorem ceases to hold in case the limiting position of a is a singular point on (II), such as a double point or cusp. See Fig. 2, § 33. While, as we shall see (§ 33), if each curve of the family has a singular point, the locus of these satisfies the geometrical as well as analytic requirements for an envelope, it is usually customary to apply the term *envelope* to that part of the locus of ultimate points of intersection of the curves of the family which has a curve of the family tangent to it at each point.

Differentiating with respect to c, we have

$$2\,cx - y = 0.$$

Eliminating c, we get $y^2 = 4\,x$, which is the singular solution.

Ex. 2. $xp^2 - 2\,yp - x = 0.$

This is the equation of Ex. 2, § 24. We saw there that its solution is

$$c^2x^2 - 2\,cy - 1 = 0.$$

Differentiating with respect to c, we have

$$cx^2 - y = 0.$$

Eliminating c, we have the singular solution

$$x^2 + y^2 = 0.$$

31. Discriminant.—If $f(z)$ is a polynomial of the nth degree,

$$c_0z^n + c_1z^{n-1} + c_2z^{n-2} + \cdots + c_{n-1}z + c_n,$$

we have by Taylor's theorem

$$f(a+h) = f(a) + f'(a)\,h + \frac{f''(a)}{2\,!}h^2 + \frac{f'''(a)}{3\,!}h^3 + \cdots + \frac{f^{(n)}(a)}{n\,!}h^n,$$

where $f'(a) = \left(\frac{df(z)}{dz}\right)_{z=a} = nc_0a^{n-1} + (n-1)\,c_1\,a^{n-2} + \cdots$

$$+ 2\,c_{n-2}a + c_{n-1},$$

$$f''(a) = \left(\frac{d^2f(z)}{dz^2}\right)_{z=a} = n(n-1)\,c_0a^{n-2} + (n-1)(n-2)c_1\,a^{n-3}$$

$$+ \cdots + 2\,c_{n-2},$$

. . . .

$$f^{(n)}(a) = \left(\frac{d^nf(z)}{dz^n}\right)_{z=a} = n\,!\,c_0.$$

Putting $h = z - a$, we have

$$f(z) = f(a) + f'(a)(z-a) + \frac{f''(a)}{2!}(z-a)^2 + \cdots + \frac{f^{(n)}(a)}{n!}(z-a)^n.$$

From this it is obvious that if a is a root of $f(z)$, *i.e.* if $f(a) = 0$, $f(z)$ contains $z - a$ as a factor; and conversely, in order that $f(z)$ should contain the factor $z - a$ we must have $f(a) = 0$. Similarly, if a is a double root, $(z-a)^2$ is a factor of $f(z)$, so that we must have $f(a) = 0$ and $f'(a) = 0$; conversely, if $f(a) = 0$ and $f'(a) = 0$, $f(z)$ will contain $(z-a)^2$ as a factor, and a will be a double root.* The necessary and sufficient condition, then, that $f(z)$ have a repeated root is that $f(z)$ and $f'(z)$ have a root, say a, or the corresponding factor $z - a$, in common. This condition obviously depends upon the coefficients of $f(z)$. That rational, entire function of the coefficients whose vanishing expresses the necessary and sufficient condition that $f(z)$ shall have a repeated root is called the *discriminant* of $f(z)$. It can readily be shown to be the product of the squares of the differences of the various roots of the equation (multiplied by a power of c_0 to avoid fractions). It may be calculated in various ways: The process of finding the greatest common divisor will show when $f(z)$ and $f'(z)$ have a common factor $z - a$. But this process is apt to introduce extraneous factors. A better way is to eliminate z from $f(z)$ and $f'(z)$, (or better still, from $nf(z) - zf'(z)$ and $f'(z)$ which are both of degree $n - 1$). The result of this elimination is a relation among the coefficients of $f(z)$, which expresses the condition that $f(z) = 0$ and $f'(z) = 0$ can be satisfied simultaneously. If all the terms of this relation are brought over to one side of the equation, and the expression is cleared of fractions and radicals, we have evidently the discriminant equated to zero. We shall call this the *discriminant relation*. Various methods for eliminating the variable

* In an entirely analogous manner it can be seen very readily that $f(a) = 0$, $f'(a) = 0$, $f''(a) = 0$, $\cdots$, $f^{(r-1)}(a) = 0$, is the necessary and sufficient condition that a be an r-fold root.

from two polynominals involving it are given in the Theory of Equations. It will be sufficient to recall here that the discriminant of the quadratic $az^2 + bz + c$ is

$$b^2 - 4\,ac,$$

while that of the cubic $az^3 + bz^2 + cz + d$ is

$$b^2c^2 + 18\,abcd - 4\,ac^3 - 4\,b^3d - 27\,a^2d^2.$$

In the case of $\phi\,(x, y, c) = 0$, looked upon as an equation in c, the coefficients are functions of x and y. It may be possible to find values of x and y, such that this equation shall have equal roots. Geometrically, this amounts to saying that there may be points through which there pass a smaller number of integral curves than usual; for to each value of c there corresponds a distinct integral curve. The discriminant relation, in this case, is the locus of such points.

32. Singular Solution Obtained directly from the Differential Equation. — Since the problem of finding the equation of the envelope of $\phi\,(x, y, c) = 0$ is identical with that of finding the discriminant relation, we see that through each point of the envelope there pass a smaller number of integral curves than through points of general position in the plane; that is, at least two of the integral curves through each point of the envelope coincide. (Thus, in the case of the family of circles already referred to, through any point of the envelope, it is readily seen, only one circle passes, instead of two.)

Since there is at least one less curve passing through a point of the envelope, there will be at least one less tangent to the curves through such a point. Hence for points along the envelope, the differential equation $f\,(p, x, y) = 0$, which defines the slopes of the tangents to the integral curves through the point (x, y), will have at least two of its roots equal, *i.e.* for points along the envelope, $f\,(p, x, y) = 0$ and $\dfrac{\partial f\,(p, x, y)}{\partial p} = 0$ are simultaneously satisfied. As a consequence, the result of eliminating p between these two equations will give us the equation of the envelope and, therefore, the singular solution, whenever there is one. See § 33.

Note. — In the previous chapter we came across certain factors, in the course of solving equations, which, while they would have led to solutions, did not contain arbitrary constants, and were therefore neglected at that time. It will now be understood that these factors usually lead to singular solutions. Thus, in the case of an equation in Clairaut's form (§ 27), (1) $y = px + f(p)$, the neglected factor is (2) $x + f'(p) = 0$. But this is exactly the derivative of (1) with respect to p. So that if we eliminate p between (1) and (2), we get the singular solution. Again, in Ex. 2 (§ 25), we neglected the factor $4p + 1 = 0$. Eliminating p between this and the original equation, we have $x + 4y = 0$, which is a singular solution of the equation, but not the whole singular solution. Both the p- and c-discriminant relations are $x(x + 4y) = 0$. This illustrates the fact that the appearance of such a factor in the course of the work implies a singular solution, but it need not always appear when a singular solution exists. In other words, this is not the way to look for singular solutions, although, in actual practice, it is advisable to examine these factors and see to what they lead.

Remark. — From the fact that two roots of an equation can be equal only in case there are as many as two roots, no singular solution can exist in the case of equations of the first order and degree. But it may, and not infrequently does, happen, that equations of a higher degree than the first have no singular solutions.

Let the student, as an exercise, prove that a differential equation of the first order and higher degree than the first, which is decomposable into factors linear in p and rational in x and y, cannot have singular solutions.

It may be further remarked that at times the singular solution gives rise to a result that is much more interesting than that arising from the general solution.

For example, let us ask for that curve which has the property of having the length of its tangent intercepted by the coördinate axes a constant l.

From formulæ (*e*), § 21, we have

$$y^2\left(\frac{p^2+1}{p^2}\right)-2xy\left(\frac{p^2+1}{p}\right)+x^2(p^2+1)=l^2,$$

or

$$(y-px)^2\left(\frac{p^2+1}{p^2}\right)=l^2.$$

Since this is in Clairaut's form (§ 27), the general solution, when solved for y, is seen at once to be

$$y=cx\pm\frac{cl}{\sqrt{c^2+1}}.$$

This represents a family of lines whose length intercepted by the axes is l. The curve that we are actually interested in is obtained when we look for the singular solution. This is gotten by finding either the p- or the c-discriminant relation (the two being identical in the case of an equation in Clairaut's form). It is

$$x^{\frac{2}{3}}+y^{\frac{2}{3}}=l^{\frac{2}{3}},$$

which is a hypocycloid of four cusps.

Ex. **1.** Find the curve for which the product of the perpendiculars drawn from two fixed points to any tangent is constant.

Ex. **2.** Find the curve whose tangents are all equidistant from the origin.

Ex. **3.** Find the curve for which the area enclosed between the tangent and the coördinate axes is a^2.

Ex. **4.** Find the curve such that the sum of the intercepts of the tangent on the coördinate axes is a constant.

Ex. 5. Integrate the following equations and examine for singular solutions:

$$x^2p^2 - 2(xy - 2)p + y^2 = 0,$$

$$(y - xp)^2 = b^2 + a^2p^2,$$

$$xp^2 - y = 0.$$

33. Extraneous Loci. — We have noticed that the c-discriminant relation is the equation of the locus of points through which a smaller number of curves pass than ordinarily. Now, if an integral curve has a double point, at that point there will be two branches of the curve. Since there are only n values of p (if the differential equation is of the nth degree) there is only room for $n - 2$ other curves through this point. Hence this point must be on the locus of the c-discriminant relation. And if there are an infinity of integral curves having double points, or nodes as they are sometimes called, the locus of these points (known as the *nodal locus*) will be given by

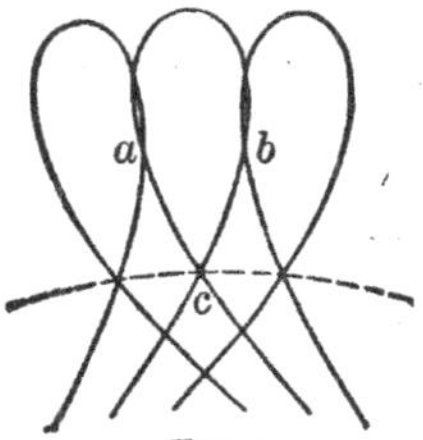

FIG. 2

FIG. 3

the c-discriminant relation. Excepting in the unusual case where this locus is also an envelope, its equation will not satisfy the differential equation. The usual case is illustrated by Fig. 2, the exceptional case by Fig. 3.

An inspection of Fig. 2 will show why the equation of the nodal locus (which, in general, is not an envelope) should be obtained when looking for the equation of the envelope. In this figure we have three neighboring curves, which coincide in the limit. The nodal locus is indicated by the broken line. Any point on it, such as c, is the limiting position of the point of intersection of the middle curve with either of the neighboring ones, *i.e.* it is the limiting position of a or b. But while in the case of the envelope, Fig. 1, § 29, a and b approach coincidence as consecutive points on the middle curve, in the case of Fig. 2, a and b approach coincidence in an entirely different way. *Consecutive points* are points which approach coincidence by moving along the same branch of a curve. In order to conclude from the theorem of § 29 that the equation of the envelope is a singular solution (§ 30), we must use the term *tangent* in the narrow sense of a line through two consecutive points. If we use it in the broader sense of a line through any two coincident points, the nodal locus may be said to be tangent to some curve of the family at each of its points.

A special case of a double point is a cusp, which may be looked upon as the limiting case of a double point, where the loop has shrunk up to the point and the two branches of the curve have become tangent. The equation of the locus of the cusps of the integral curves, known as the *cuspidal locus*, found when the equation of the envelope is sought, will be a solution only in case this locus is also an envelope (as in case of Fig. 4). Otherwise, it is not a solution (as in Fig. 5).

In the case of a cusp, not only is the number of integral curves through that point at least one less than the usual number, that is, not only does the c-discriminant vanish at this point, but two values of p are equal there, since the tangents to the two branches of the curve coincide, that is, at such a point the p-discriminant also vanishes.

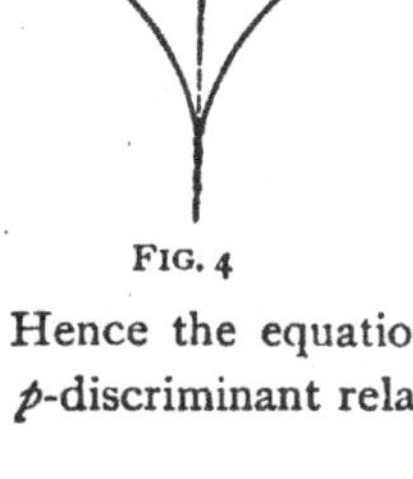

FIG. 4

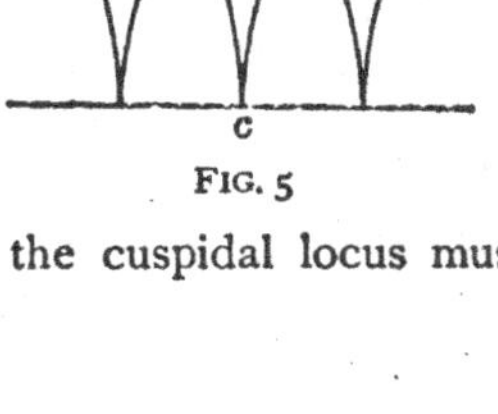

FIG. 5

Hence the equation of the cuspidal locus must also appear in the p-discriminant relation.

So far, these extraneous loci, which may or may not be solutions, have arisen as results of peculiarities of the integral curves. Thus, if the integral curves are known to have no double points or cusps, it is clear there can be no nodal or cuspidal loci. But an extraneous locus may arise, irrespective of the character of the integral curves. Wherever two distinct integral curves are tangent to each other, while the number of curves through that point is unaffected, the number of distinct values of p is diminished. Hence the p-discriminant vanishes at that point, and the locus of such points, if it exists, will be given by the p-discriminant relation. This locus is known as the *tac-locus*, and its equation may or may not satisfy the differential equation. Thus in the case of the family of circles referred to in the footnote, page 61, the p-discriminant relation is found to be $y^2(y^2-r^2)=0$; here $y=\pm r$ is the envelope, while $y=0$ is the tac-locus. By actual trial, it is found that $y=0$ does not satisfy the differential equation.

Remark.—At times, as the parameter approaches a certain value, the curves of the family approach a limiting one, usually different in shape from all the others. Frequently this special curve of the family has the property of being tangent to all the others at one point. Excepting at this point (through which there is an infinite number of curves), a smaller number of curves than usual pass through every point of the special curve. Hence the equation of the latter is given by both the p- and c-discriminant relations. Moreover, it is found that the factor corresponding to this special solution appears once in the c-discriminant and three times in the p-discriminant.

Thus, in the case of Ex. 4, § 24, the integral curves are a family of cubics tangent to the axis of y at the origin (see Ex. 8, § 20). Their equation is $y^2 = 2x(x-c)^2$. For $c=\infty$, we have the curve $x=0$, which is tangent to every other one at the origin. The c- and p-discriminants are respectively xy^2 and x^5. The additional factor x^2 appears because $x=0$ is also a tac-locus, it being a particular solution corresponding to the two distinct values $c=\pm\infty$.

Besides, in 1888, Mr. J. M. Hill* proved that the factors in the c-discriminant corresponding to the envelope, nodal, and cuspidal locus occur once, twice, and three times respectively, while those corresponding to the envelope, tac-locus and cuspidal locus in the p-discriminant occur once, twice, and once respectively. All this can be put in tabular form, as follows:

c-discriminant		p-discriminant	
envelope	1	envelope	1
particular curve	1	particular curve	3
nodal locus	2	tac-locus	2
cuspidal locus	3	cuspidal locus	1

In case any locus comes under two heads, the factor corresponding will occur the number of times it should for each of the heads; thus, if the tac-locus is also an envelope, that factor will occur once in the c-discriminant and three times in the p-discriminant; and if the cuspidal locus is an envelope, it will occur four times in the c-discriminant and twice in the p-discriminant, and so on.

This will often prove a check on the work, although it should be only relied upon as a check and not as the only clew to identify the results of finding the discriminants. The process of finding discriminants is frequently beset with chances to introduce or to drop a factor, so that, unless great care is taken, the number of times a factor is found to occur may not be the correct one, and inferences drawn from it will be false. In actual practice it is desirable to find both the p- and the c-discriminants, and then test their various factors, equated to zero, to see if they satisfy the equation. In this way, if a factor has been lost in either one of the discriminants, its appearance in the other will keep it from being lost as a solution, while a blind use of the table would cause one to give a different interpretation to the result.

* *Proc. Lond. Math. Society*, Vol. XIX, p. 561.

In certain cases there can be no doubt. Thus, if the degree of the equation is two or three, the use of the formulæ mentioned in § 31 will give all the factors occurring the correct number of times. Again, in case the integral curves are straight lines (as is always the case when the equation is in Clairaut's form), there is no need of looking for any of the extraneous loci.

Again, if the integral curves are conics, there can be no nodal or cuspidal loci.

Examine the following equations for singular solutions and extraneous loci :

Ex. 1. $xp^2 - (x - 1)^2 = 0.$

The general solution is readily seen to be

$$9(y+c)^2 = 4x(x-3)^2,$$

which is the equation of a family of nodal cubics, each of which is tangent to the axis of y and has its node at the point $(3, c)$.

The p-discriminant relation is $x(x-1)^2 = 0$, while the c-discriminant relation is $x(x-3)^2 = 0$.

Here $x = 0$ is common to the two. It also satisfies the equation. [For the line $x = 0$, $p = \infty$ at every point.] Hence it is the singular solution.

$x - 1 = 0$ occurs in the p-discriminant only. It is the tac-locus. [Notice that this factor occurs twice.]

$x - 3 = 0$ occurs in the c-discriminant only. It is the nodal locus. [Notice that this factor occurs twice.]

Ex. 2. $8(1+p)^3 = 27(x+y)(1-p)^3.$

The general solution is

$$(x-y+c)^3 = (x+y)^2.$$

As it is rather awkward to substitute the coefficients in the formula for the discriminant given in § 31, make the substitution

$$x+y=\xi, \quad x-y=\eta.$$

The equation then becomes

$$27\,\xi\left(\frac{d\eta}{d\xi}\right)^3=8,$$

and the solution becomes $\qquad (\eta+c)^3=\xi^2.$

Now the p-discriminant relation is $\xi^2=0$, and the c-discriminant relation is $\xi^4=0$.

$\xi=0$ is common to both, and satisfies the equation. Hence $\xi=0$, or $x+y=0$, is a singular solution.

It is also a cuspidal locus, as may be seen by constructing some of the semicubical parabolas $(\eta+c)^3=\xi^2$. (See Fig. 4.) Note the number of times that these factors occur.

Ex. 3. $4p^2=9x$.

The general solution is

$$(y+c)^2=x^3.$$

Here the p-discriminant relation is $x=0$, and the c-discriminant relation is $x^3=0$.

It is obvious that $x=0$ does not satisfy the equation. It is a cuspidal locus.

Ex. 4. Examine the following equations for singular solutions and extraneous loci:

$y(3-4y)^2p^2=4(1-y)$.

§ 24, Ex. 3, 4, 5. § 25, Ex. 5, 6. § 26, Ex. 2, 4. § 27, Ex. 2, 6. § 28, Ex. 1, 2, 3, 5, 11.

Ex. 5. The family of circles determined by Ex. 5, § 26, envelops a curve whose equation is a singular solution of the differential equation. Find it.

34. Summary. — We have seen that the equation of the singular solution (or of the envelope) is given by both the c- and p-discriminant relations (§§ 30, 32). Moreover, the c-discriminant relation gives rise to the nodal and cuspidal loci, while the p-discriminant relation gives rise to the cuspidal and tac-loci, while both of them, at times, give rise to a particular solution § 33. For the number of times the corresponding factors occur in each discriminant, see remark, § 33.*

Remark. — It should be noted that, in general, a differential equation has no singular solution. For $f(p, x, y) = 0$ and $\frac{\partial f}{\partial p} = 0$ can be solved for y and p giving

$$y = \phi(x), \quad p = \phi_1(x).$$

In order that this value of y be a solution we must have

$$\phi_1(x) = \frac{d\phi(x)}{dx},$$

which is not true, in general. Darboux proved that, in general, the result of eliminating p from the above two equations is the equation of the cuspidal locus. (*Bulletin des Sciences Mathématiques*, 1873, p. 158.) Picard also gives a proof of this in his *Traité d'Analyse*, Vol. III, p. 45. See also Fine's article in the *American Journal of Mathematics*, Vol. XII; Chrystal, *Nature*, 1896; Liebmann, *Differentialgleichungen*, p. 95. This may seem at first sight contrary to what is to be expected from the way in which the idea of a singular solution was introduced. (It was Lagrange (1736–1813) who first noted that the equation of the envelope of the family of integral curves is a solution.) But it has already been noted that in the process of finding the equation of the envelope, extraneous loci may arise, and it turns out that these usually do arise to the exclusion of an envelope (see Picard, Vol. III, p. 51). Moreover, all this was based on the assumption that the general solution of the equation has the form (1), where c enters rationally. While this is true in a very large class of equations, it is nevertheless only a special case.

* The theory as given here was first developed by Arthur Cayley (1821–95), *Messenger of Mathematics*, Vol. II (1872), p. 6, Vol. VI, p. 23. For illustrative examples see J. W. L. Glaisher, *Messenger of Mathematics*, Vol. XII, p. 1.

CHAPTER VI

TOTAL DIFFERENTIAL EQUATIONS *

35. Total Differential Equations. A differential equation, involving three or more variables, of the form

$$(1) \qquad P(x, y, z)\, dx + Q(x, y, z)\, dy + R(x, y, z)\, dz = 0$$

is called a *total differential equation.* We shall consider the case when its solution can be put in the form

$$u(x, y, z) = c.$$

The differential equation arising from this is

$$(2) \qquad \frac{\partial u}{\partial x} dx + \frac{\partial u}{\partial y} dy + \frac{\partial u}{\partial z} dz = 0.$$

This is either the same as (1), or differs from it by a factor $\mu(x, y, z)$; *i. e.* if (1) is integrable, there must be an integrating factor for it. Then a function $\mu(x, y, z)$ exists, such that

$$\frac{\partial u}{\partial x} = \mu P, \quad \frac{\partial u}{\partial y} = \mu Q, \quad \frac{\partial u}{\partial z} = \mu R.$$

Hence

$$\mu \frac{\partial Q}{\partial z} + Q \frac{\partial \mu}{\partial z} = \mu \frac{\partial R}{\partial y} + R \frac{\partial \mu}{\partial y}, \qquad \text{since } \frac{\partial^2 u}{\partial y\, \partial z} = \frac{\partial^2 u}{\partial z\, \partial y},$$

* For certain reasons it seems desirable to consider this class of equations before going to the study of differential equations of higher order than the first. If desired, this chapter may be taken up after Chapter IX.

$$\mu\frac{\partial R}{\partial x}+R\frac{\partial \mu}{\partial x}=\mu\frac{\partial P}{\partial z}+P\frac{\partial \mu}{\partial z}, \qquad \text{since } \frac{\partial^2 u}{\partial z\,\partial x}=\frac{\partial^2 u}{\partial x\,\partial z},$$

$$\mu\frac{\partial P}{\partial y}+P\frac{\partial \mu}{\partial y}=\mu\frac{\partial Q}{\partial x}+Q\frac{\partial \mu}{\partial x}, \qquad \text{since } \frac{\partial^2 u}{\partial x\,\partial y}=\frac{\partial^2 u}{\partial y\,\partial x}.^*$$

μ must satisfy these three equations, which of course cannot be expected of it, unless P, Q, R satisfy a certain condition or conditions.

If we multiply these equations by P, Q, R respectively and add, all of the derivatives of μ disappear, and we have left

$$(3) \qquad P\left(\frac{\partial Q}{\partial z}-\frac{\partial R}{\partial y}\right)+Q\left(\frac{\partial R}{\partial x}-\frac{\partial P}{\partial z}\right)+R\left(\frac{\partial P}{\partial y}-\frac{\partial Q}{\partial x}\right)=0,^\dagger$$

after dropping the common factor μ, which is not zero, for the introduction of zero as an integrating factor gives us no information.

This is a necessary condition among the coefficients P, Q, R. Moreover, we shall prove that it is the only condition requisite for the existence of an integral of (1), in other words, we shall prove that this condition is also sufficient.

Consider any one of the variables, say z, as a constant temporarily. Then equation (1) takes the form

$$(4) \qquad P\,dx+Q\,dy=0.$$

* Assuming the continuity of μP, μQ, μR, and the existence and continuity of their derivatives.

† This may be written in the following symbolic determinant form:

$$\begin{vmatrix} P & Q & R \\ \frac{\partial}{\partial x} & \frac{\partial}{\partial y} & \frac{\partial}{\partial z} \\ P & Q & R \end{vmatrix}=0,$$

which is very easy to retain in mind.

Still considering z as constant, (4) can be integrated, but now the constant of integration may involve z. Let the solution be

(5) $$u(x, y, z) = \phi(z).$$

We shall show that if the condition (3) is satisfied, we can choose $\phi(z)$ so that (5) will be the solution of (1). For, differentiating (5) with respect to all the variables, we have

(6) $$\frac{\partial u}{\partial x}dx + \frac{\partial u}{\partial y}dy + \frac{\partial u}{\partial z}dz = d\phi.$$

Since (5) is a solution of (4) considering z as constant, we have

(7) $$\frac{\partial u}{\partial x} = \mu(x, y, z)P, \quad \frac{\partial u}{\partial y} = \mu(x, y, z)\,Q,$$

where μ is, in fact, an integrating factor of (4). (See § 5.)

Comparing (6) with (1) multiplied by μ, we have

(8) $$\left(\frac{\partial u}{\partial z} - \mu R\right)dz = d\phi.$$

This equation can be solved for ϕ provided $\frac{\partial u}{\partial z} - \mu R$ reduces to a function of z and ϕ, when use is made of (5), or, what is the same thing, provided $\frac{\partial u}{\partial z} - \mu R$ is a function of z and u. Looking upon $\frac{\partial u}{\partial z} - \mu R$ and u as functions of x and y only, z being treated as a constant or parameter, the only requirement for this is the vanishing of their Jacobian: *

$$\begin{vmatrix} \dfrac{\partial u}{\partial x}, & \dfrac{\partial^2 u}{\partial x\,\partial z} - \mu\dfrac{\partial R}{\partial x} - R\dfrac{\partial \mu}{\partial x} \\ \dfrac{\partial u}{\partial y}, & \dfrac{\partial^2 u}{\partial y\,\partial z} - \mu\dfrac{\partial R}{\partial y} - R\dfrac{\partial \mu}{\partial y} \end{vmatrix}.$$

* See Note I in the Appendix.

Making use of (7), this may be written

$$\begin{vmatrix} \mu P, & \mu\dfrac{\partial P}{\partial z}+P\dfrac{\partial \mu}{\partial z}-\mu\dfrac{\partial R}{\partial x}-R\dfrac{\partial \mu}{\partial x} \\ \mu Q, & \mu\dfrac{\partial Q}{\partial z}+Q\dfrac{\partial \mu}{\partial z}-\mu\dfrac{\partial R}{\partial y}-R\dfrac{\partial \mu}{\partial y} \end{vmatrix},$$

or $$\mu^2\left[P\left(\frac{\partial Q}{\partial z}-\frac{\partial R}{\partial y}\right)+Q\left(\frac{\partial R}{\partial x}-\frac{\partial P}{\partial z}\right)\right]-\mu R\left(P\frac{\partial \mu}{\partial y}-Q\frac{\partial \mu}{\partial x}\right).$$

Assuming that (3) is satisfied, this becomes

$$\mu R\left[\mu\left(\frac{\partial Q}{\partial x}-\frac{\partial P}{\partial y}\right)+Q\frac{\partial \mu}{\partial x}-P\frac{\partial \mu}{\partial y}\right],$$

or $$\mu R\left[\frac{\partial(\mu Q)}{\partial x}-\frac{\partial(\mu P)}{\partial y}\right].$$

Since μ is an integrating factor of (4), this vanishes. Hence equation (8) can be solved for ϕ; putting this in (5), we have the solution of our equation (1), and our theorem is proved.*

* The necessity and sufficiency of the condition (3) can be proved much more briefly as follows (but this method does not suggest a general way of solving a total differential equation when the condition is satisfied):

The equation $$P\,dx+Q\,dy+R\,dz=0$$

is equivalent to the two partial differential equations,

$$\frac{\partial z}{\partial x}=-\frac{P}{R},\quad \frac{\partial z}{\partial y}=-\frac{Q}{R}.$$

In order that these may hold simultaneously, it is necessary and sufficient that

$$\frac{\partial\left(\frac{P}{R}\right)}{\partial y}=\frac{\partial\left(\frac{Q}{R}\right)}{\partial x}.$$

Remembering that P, Q, R are functions of x, y, z, this equation becomes

$$R\left(\frac{\partial P}{\partial y}+\frac{\partial P}{\partial z}\frac{\partial z}{\partial y}\right)-P\left(\frac{\partial R}{\partial y}+\frac{\partial R}{\partial z}\frac{\partial z}{\partial y}\right)=R\left(\frac{\partial Q}{\partial x}+\frac{\partial Q}{\partial z}\frac{\partial z}{\partial x}\right)-Q\left(\frac{\partial R}{\partial x}+\frac{\partial R}{\partial z}\frac{\partial z}{\partial x}\right).$$

Since $\dfrac{\partial z}{\partial x}=-\dfrac{P}{R}$, $\dfrac{\partial z}{\partial y}=-\dfrac{Q}{R}$, this reduces at once to the form (3) above.

36. Method of Solution. — The above proof not only establishes the sufficiency of the condition, it also suggests the following method for solving a total differential equation in three variables which satisfies this condition:

Integrate the equation considering one of the variables * *as a constant. Instead of a constant of integration, introduce an undetermined function of this variable. Redifferentiate, this time with respect to all the variables. Comparing this with the original differential equation, a new differential equation will arise, involving only the undetermined function and that variable of which it is a function. From this the function can be determined, involving an arbitrary constant. And thus the complete solution is found.*

Remark. — Since an equation which is integrable differs only by a factor from an exact differential equation, if we can obtain such a factor by inspection or otherwise, we can integrate at once.

Apply test for integrability and integrate the following:

Ex. 1. $y^2\,dx + z\,dy - y\,dz = 0.$

$$\begin{vmatrix} y^2 & z & -y \\ \frac{\partial}{\partial x} & \frac{\partial}{\partial y} & \frac{\partial}{\partial z} \\ y^2 & z & -y \end{vmatrix} = y^2(-1-1) + z(0-0) - y(0-2y) = -2y^2 + 2y^2 = 0.$$

While in this simple case there is very little choice in selecting the variable to be constant, there is, perhaps, a little advantage in letting y be so chosen. Then we have

$$y^2\,dx - y\,dz = 0, \quad \text{whence } yx - z = \phi(y).$$

Differentiating,
$$y\,dx + x\,dy - dz = d\phi,$$

or
$$y^2\,dx + xy\,dy - y\,dz = y\,d\phi.$$

* We usually choose that variable a constant which will have the effect of simplifying the resulting equation in the other two variables as much as possible.

Comparing this with the differential equation, we have

$$(xy - z)\,dy = y\,d\phi,$$

or

$$\phi\,dy = y\,d\phi.$$

$$\therefore\ \phi = cy.$$

Hence the general solution is $yx - z + cy = 0$.

By inspection, it is readily noticeable that $\frac{1}{y^2}$ is an integrating factor. This puts the equation in the form

$$dx + \frac{zdy - ydz}{y^2} = 0.$$

Its solution is of course $x - \frac{z}{y} + c = 0$.

Ex. 2. $zy\,dx - zx\,dy - y^2\,dz = 0$.

Ex. 3. $x\,dx + y\,dy - \sqrt{a^2 - x^2 - y^2}\,dz = 0$.

Ex. 4. $(x^2 - y^2 - z^2)dx + 2\,xy\,dy + 2\,xz\,dz = 0$.

37. Homogeneous Equations. — If P, Q, and R are homogeneous and of the same degree, the variables may be separated just as in the corresponding case for two variables (§ 10). Here we transform any two of the variables, say x and y, by $x = uz$, $y = vz$. Then $dx = z\,du + u\,dz$, $dy = z\,dv + v\,dz$, and the differential equation becomes

$$z(P_1\,du + Q_1\,dv) + (uP_1 + vQ_1 + R_1)dz = 0,\text{*}$$ or

(1) $$\frac{P_1\,du + Q_1\,dv}{uP_1 + vQ_1 + R_1} + \frac{dz}{z} = 0,$$

where $P_1 = P(u, v, 1)$, $Q_1 = Q(u, v, 1)$, $R_1 = R(u, v, 1)$.

* If $uP_1 + vQ_1 + R_1 = 0$, this equation reduces at once to one in the two variables u and v.

Now, if the original equation satisfies the condition for integrability, this equation will also. Moreover, it is exact* and can be integrated as it stands (by method of § 8).

Ex. 1. $(y^2+yz)dx+(xz+z^2)dy+(y^2-xy)dz=0.$

(Let the student apply the test for integrability.)

Putting $x=uz$, $y=vz$, $dx=u\,dz+z\,du$, $dy=v\,dz+z\,dv$, and the equation becomes

$$\frac{dz}{z}+\frac{(v^2+v)\,du+(u+1)\,dv}{uv^2+uv+v+v^2}=0.$$

Since $uv^2+uv+v+v^2=(v^2+v)(u+1)$, the equation may be written

$$\frac{dz}{z}+\frac{du}{u+1}+\frac{dv}{v^2+v}=0;$$

whence

$$\frac{z(u+1)v}{v+1}=c,$$

or

$$\frac{y(x+z)}{y+z}=c.$$

Ex. 2. $(y^2+yz+z^2)dx+(z^2+zx+x^2)dy+(x^2+xy+y^2)dz=0.$

Ex. 3. $(x^2y-y^3-y^2z)dx+(xy^2-x^2z-x^3)dy+(xy^2+x^2y)dz=0.$

* Putting $\bar{P}=\dfrac{P_1}{uP_1+vQ_1+R_1}$, $\bar{Q}=\dfrac{Q_1}{uP_1+vQ_1+R_1}$, $\bar{R}=\dfrac{1}{z}$,

equation (1) takes the form $\bar{P}\,du+\bar{Q}\,dv+\bar{R}\,dz=0$.

Since $\bar{P}$ and $\bar{Q}$ are free of z, and $\bar{R}$ is free of u and v, the condition for integrability reduces to

$$\bar{R}\left(\frac{\partial\bar{Q}}{\partial u}-\frac{\partial\bar{P}}{\partial v}\right)=0.$$

$\bar{R}$ is $\frac{1}{z}$, hence we must have $\frac{\partial\bar{Q}}{\partial u}-\frac{\partial\bar{P}}{\partial v}=0$, which means that $\bar{P}\,du+\bar{Q}\,dv$ is an exact differential (§ 7), and therefore (1) is also.

38. Equations involving more than Three Variables. — Consider the equation

(1) $$Pdx + Q\,dy + R\,dz + S\,dt = 0.$$

If this is integrable, it will remain so when we let any of the variables be a constant. Letting x, y, z, t be constants successively, the conditions are

(2) $$Q\left(\frac{\partial S}{\partial z} - \frac{\partial R}{\partial t}\right) + R\left(\frac{\partial Q}{\partial t} - \frac{\partial S}{\partial y}\right) + S\left(\frac{\partial R}{\partial y} - \frac{\partial Q}{\partial z}\right) = 0,$$

(3) $$R\left(\frac{\partial P}{\partial t} - \frac{\partial S}{\partial x}\right) + S\left(\frac{\partial R}{\partial x} - \frac{\partial P}{\partial z}\right) + P\left(\frac{\partial S}{\partial z} - \frac{\partial R}{\partial t}\right) = 0,$$

(4) $$S\left(\frac{\partial Q}{\partial x} - \frac{\partial P}{\partial y}\right) + P\left(\frac{\partial S}{\partial y} - \frac{\partial Q}{\partial t}\right) + Q\left(\frac{\partial P}{\partial t} - \frac{\partial S}{\partial x}\right) = 0,$$

(5) $$P\left(\frac{\partial R}{\partial y} - \frac{\partial Q}{\partial z}\right) + Q\left(\frac{\partial P}{\partial z} - \frac{\partial R}{\partial x}\right) + R\left(\frac{\partial Q}{\partial x} - \frac{\partial P}{\partial y}\right) = 0.*$$

But these conditions are not all independent. If we multiply (2), (3), (4) by P, Q, R respectively and add, we get (5) multiplied by S; which shows that only the first three are independent.†

* Perfectly generally, if the equation contains n variables, we obtain as many conditions as the number of ways in which we can pick out three variables from n; that is, the number of conditions is $\frac{n(n-1)(n-2)}{3!}$.

† In general, the number of independent conditions in the case of n variables is $\frac{(n-1)(n-2)}{2}$, which is the number of times two objects can be chosen out of $n-1$. For only those conditions will be independent which involve derivatives with respect to some one chosen variable, since any condition not involving such can be obtained by combining linearly those that do, as was done in the case above. Now each of the conditions involves derivatives with respect to three variables. Hence the derivatives with respect to any one variable may appear in a condition along with those with respect to any two of the remaining $n-1$ variables.

These conditions can also be shown to be sufficient. When they hold, the integral is found as in the case of three variables, by integrating, considering all but two of the variables constant. Then the constant of integration is written as a function ϕ of those variables temporarily considered constant. Redifferentiating with respect to all the variables, and comparing with the given equation, the two variables originally treated differently from the rest will disappear, and the function will enter in a new differential equation which is integrable and involves the remaining $n-2$ variables and ϕ, that is $n-1$ variables in all. Either this can be integrated at once, or the process may be repeated as often as necessary. The following example will illustrate.

Ex. $z(y+z)dx + z(t-x)dy + y(x-t)dz + y(y+z)dt = 0.$

Let y and z be constants temporarily. Integrating, we have

$$xz + yt = \phi(y, z).$$

Differentiating and comparing with the original equation, we have

$$(ty + zx)(dy + dz) = (y+z)d\phi.$$

or

$$\phi\,(dy + dz) = (y+z)\,d\phi.$$

We now have an equation in the three variables y, z, ϕ. This can be solved by the general method (§ 36). But an obvious integrating factor is $\frac{1}{(y+z)\,\phi}$. Introducing this, we have

$$\phi = c(y+z).$$

Hence the general solution is

$$xz + yt = c(y+z).$$

39. Equations which do not satisfy the Condition for Integrability. If $P\,dx + Q\,dy + R\,dz = 0$ does not satisfy the condition for integrability, it is impossible to find its general solution in the form

$$\phi(x, y, z) = c.$$

But as the equation is one in three variables, we should expect to find an indefinite number of solutions. As a matter of fact, if we assume any relation we please, say $\psi(x, y, z) = 0$, this will determine any one of the variables, say z, in terms of the other two. Substituting for this variable in the original equation, we obtain a new differential equation in two variables, which can usually be solved. We see then that the general solution of a so-called *non-integrable* total differential equation consists of an arbitrarily chosen relation among the variables and a second relation involving an arbitrary constant. The latter depends upon the choice of the former, and cannot be determined until the choice has been made.

Remark. — Since the solution of the integrable equation is a single relation among the three variables, we may assume any second one consistent with it. So that in this case also we may say that the solution consists of an arbitrarily chosen relation and a second one involving an arbitrary constant. But here the latter is fixed by the differential equation, and is entirely independent of the choice of the former.

Ex. $y\,dx + x\,dy - (x + y + z)\,dz = 0.$

This, it is readily seen, does not satisfy the condition for integrability. If we assume $x + y + z = 0$, our equation becomes

$$y\,dx + x\,dy = 0, \text{ whose solution is } xy = c.$$

Hence a solution is $\begin{cases} x + y + z = 0, \\ xy = c. \end{cases}$

If we assume $x + y = 0$, our equation becomes $y\,dx + x\,dy - z\,dz = 0$, whose solution is $2\,xy - z^2 = c$. Hence another solution is

$$\begin{cases} x + y = 0, \\ 2\,xy - z^2 = c. \end{cases}$$

40. Geometrical Interpretation. — To say that the equation

$$P\,dx + Q\,dy + R\,dz = 0$$

satisfies the condition for integrability is to say that a family of surfaces

$$\phi(x, y, z) = c$$

exists such that at each point (x_0, y_0, z_0) in space there passes one* of these surfaces

$$\phi(x, y, z) = \phi(x_0, y_0, z_0),$$

and the tangent plane at any point (x, y, z) of this surface is

$$P(x,y,z)(X - x) + Q(x,y,z)(Y - y) + R(x,y,z)(Z - z) = 0$$

In other words, the differential equation defines the plane $P(X-x)+Q(Y-y)+R(Z-z)=0$ at each point in space. The problem of integration amounts to determining a family of surfaces† such that the surface which passes through any point is tangent to the plane corresponding to that point. An interesting consequence of this is brought out in § 66.

When the equation is not integrable, the assumption of a second relation, $\psi(x, y, z) = 0$, which carries with it $\frac{\partial\psi}{\partial x}dx + \frac{\partial\psi}{\partial y}dy + \frac{\partial\psi}{\partial z}dz = 0$, determines, with the original equation, a line at each point on the assumed surface $\psi(x, y, z) = 0$, viz.

$$\left\{\begin{aligned} &P(X-x) + Q(Y-y) + R(Z-z) = 0, \\ &\frac{\partial\psi}{\partial x}(X-x) + \frac{\partial\psi}{\partial y}(Y-y) + \frac{\partial\psi}{\partial z}(Z-z) = 0. \end{aligned}\right\}$$

The problem of integration then amounts to determining a family of curves such that that curve which passes through any point is tangent to the line corresponding to that point. Since one of the two

* It is presupposed here that ϕ is a rational function of x, y, z. Otherwise the statement in the text must be restricted to regions in which ϕ is single-valued. (See § 70.)

† These will be referred to as *integral surfaces.*

equations of this family of curves is the assumed relation $\psi(x, y, z) = 0$, the problem really amounts to finding that family of curves on any arbitrarily chosen surface whose tangent at any point of the surface lies in the plane determined by the differential equation at that point. Thus, in the case of Ex., § 39, we have on the plane $x + y + z = 0$ the curves cut out upon it by the family of cylinders $xy = c$; while on the plane $x + y = 0$ we have the curves cut out by the hyperbolic paraboloids $2xy - z^2 = c$.*

41. Summary. — If the total differential equation

$$P\,dx + Q\,dy + R\,dz = 0$$

satisfies the condition for integrability (§ 35), an integrating factor exists. If that can be found by inspection, introduce it, and integrate at once.

If the integrating factor cannot be found by inspection, the general method of § 36 may be employed.

If P, Q, R are homogeneous and of the same degree, the method of § 37 will sometimes prove simpler than the general method.

If the condition for integrability is not satisfied, solutions may be found by the method of § 39.

Total differential equations involving more than three variables may be treated by the method of § 38, unless an integrating factor is obvious by inspection. In this case introduce the factor and integrate at once.

Apply the test for integrability, and solve the following:

Ex. 1. $(y + z)dx + (z + x)dy + (x + y)dz = 0$.

Ex. 2. $(z + 1)(x\,dx + y\,dy) - (x^2 + y^2)dz = 0$.

Ex. 3. $(x + y^2 + z^2 + 1)dx + 2y\,dy + 2z\,dz = 0$.

Ex. 4. $(y + a)^2\,dx + z\,dy - (y + a)dz = 0$.

* All this applies to integrable equations, except that in case the arbitrarily chosen surface is an integral surface, every curve on it is an integral curve.

Ex. **5.** $(y+z)dx+dy+dz=0.$

Ex. **6.** $2x\,dx+dy+(2x^2z+2yz+2z^2+1)dz=0.$

Ex. **7.** $(2x+y^2+2xz)dx+2xy\,dy+x^2\,dz-dt=0.$

Ex. **8.** $zx\,dy-yz\,dx+x^2\,dz=0.$

Ex. **9.** $x(y-1)(z-1)dx+y(z-1)(x-1)dy$
$+z(x-1)(y-1)dz=0.$

Ex. **10.** $(y-z)dx+2(x+3y-z)dy-2(x+2y)dz=0.$

Ex. **11.** $t(y+z)dx+t(y+z+1)dy+t\,dz-(y+z)dt=0.$

Ex. **12.** $z(y+z)dx+z(t-x)dy+y(x-t)dz+y(y+z)dt=0.$

CHAPTER VII

LINEAR DIFFERENTIAL EQUATIONS WITH CONSTANT COEFFICIENTS

42. General Linear Differential Equation. — A *linear differential equation* is one which is of the first degree in the dependent variable and all of its derivatives. Its general type is

$$(1) \qquad X_0\frac{d^n y}{dx^n} + X_1\frac{d^{n-1}y}{dx^{n-1}} + X_2\frac{d^{n-2}y}{dx^{n-2}} + \cdots + X_{n-1}\frac{dy}{dx} + X_n y = X,$$

where $X_0, X_1, X_2, \cdots, X_n, X$ are functions of x or constants. If we write $\frac{dy}{dx} = Dy, \frac{d^2y}{dx^2} = D^2y, \cdots, \frac{d^ny}{dx^n} = D^ny$, we can write (1) in the following convenient form,

$$(X_0 D^n + X_1 D^{n-1} + X_2 D^{n-2} + \cdots + X_{n-1} D + X_n) y = X,$$

or

$$F(D) y = X,$$

where $F(D)$ is the polynomial $X_0 D^n + X_1 D^{n-1} + \cdots + X_n$ which represents symbolically the differential operator

$$X_0\frac{d^n}{dx^n} + X_1\frac{d^{n-1}}{dx^{n-1}} + \cdots + X_{n-1}\frac{d}{dx} + X_n.$$

Two properties of linear differential equations which are of service in their solution deserve especial mention here:

1° Suppose $X = 0$. In this case the equation is said to be a *homogeneous* linear differential equation, since all of its terms are of the first degree in y and its derivatives. (When not homogeneous, the equation is said to be a *complete* linear differential equation.) If $y = y_1$ satisfies the equation, so will $y = c_1 y_1$, where c_1 is a constant. For, since $D^k(c_1 y_1) = c_1 D^k y_1$, $F(D)(c_1 y_1) = c_1 F(D) y_1$. But, by hypothesis, $F(D) y_1 = 0$, hence $F(D)(c_1 y_1) = 0$.

Moreover, if $y = y_2$ is also a solution, $y = c_1 y_1 + c_2 y_2$ will be a solution. For, since the derivative of the sum is the sum of the derivatives, *i.e.* $D^k(y_1 + y_2) = D^k y_1 + D^k y_2$, we have

$$F(D)(c_1 y_1 + c_2 y_2) = F(D)(c_1 y_1) + F(D)(c_2 y_2)$$
$$= c_1 F(D) y_1 + c_2 F(D) y_2 = 0.$$

Similarly, if we know r particular integrals, $y_1, y_2, \cdots, y_r$,

$$y = c_1 y_1 + c_2 y_2 + \cdots + c_r y_r$$

will be a solution. Since the general solution of a differential equation of the nth order is a solution which involves n independent arbitrary constants, we have the property:

A. If $y_1, y_2, \cdots, y_n$ are n linearly independent particular integrals of a homogeneous linear differential equation of the nth order, the function $c_1 y_1 + c_2 y_2 + \cdots + c_n y_n$ is its general integral.*

If the particular integrals are not linearly independent, the solution found above will not be the general solution. Thus, suppose there exists the relation $a_1 y_1 + a_2 y_2 + \cdots + a_n y_n \equiv 0$, where all the a's are not zero. If a_n is different from zero, $y_n \equiv -\frac{a_1}{a_n} y_1 - \frac{a_2}{a_n} y_2 - \cdots - \frac{a_{n-1}}{a_n} y_{n-1}$, and the integral becomes

$$\left(c_1 - \frac{a_1}{a_n}\right) y_1 + \left(c_2 - \frac{a_2}{a_n}\right) y_2 + \cdots + \left(c_{n-1} - \frac{a_{n-1}}{a_n}\right) y_{n-1},$$

where only $n - 1$ independent constants are involved.

* n functions $y_1, y_2, \cdots, y_n$ of a variable are said to be *linearly independent* if it is impossible to find n constants $a_1, a_2, \cdots, a_n$ such that $a_1 y_1 + a_2 y_2 + \cdots + a_n y_n$ shall vanish for all values of the variables. Thus $y_1 = 2x - x^2$, $y_2 = x + x^2$, $y_3 = x$ are evidently not linearly independent, since $y_1 + y_2 - 3y_3 \equiv 0$; *i.e.* it equals zero for all values of x, or, as it is usually expressed, $y_1 + y_2 - 3y_3$ vanishes identically.

Remark.—Attention should be called to the fact that it makes no difference how these particular integrals are gotten. We shall see that in a most commonly occurring class of equations, these will be found by purely algebraic means; in other cases, some of them can be gotten by inspection.

For convenience of language, the integral of (1), when the right-hand member is made zero temporarily, is spoken of as the *complementary function.*

2° If $Y \equiv c_1 y_1 + c_2 y_2 + \cdots + c_n y_n$ is the complementary function of (1), and if we know (no matter by what means) a particular integral, U, then $Y + U$ is the general integral of (1). For, since the equation is linear,

$$f(D)(Y+U) = f(D)\,Y + f(D)\,U = 0 + X = X.$$

Hence the property:

B. The general integral of a complete linear differential equation is the sum of its complementary function and any particular integral.

43. Linear Differential Equations with Constant Coefficients.* Complementary Function.—Given the equation

$$(1) \qquad k_0 \frac{d^n y}{dx^n} + k_1 \frac{d^{n-1} y}{dx^{n-1}} + k_2 \frac{d^{n-2} y}{dx^{n-2}} + \cdots + k_{n-1} \frac{dy}{dx} + k_n y = X,$$

or

$$(k_0 D^n + k_1 D^{n-1} + k_2 D^{n-2} + \cdots + k_{n-1} D + k_n) y = X,$$

or

$$f(D) y = X,$$

where $k_0, k_1, \cdots, k_n$ are constants.

First, suppose $X = 0$. Then

$$(2) \qquad f(D) y = 0.$$

* The method given here is due to Leonhard Euler (1707–1783). For a presentation of Cauchy's method see T. Craig, *A Treatise on Linear Differential Equations*, Vol. I, Ch. II; or C. Hermite, "Équations Différentielles Linéaires," in *Bulletin des Sciences Mathématiques*, 1879.

Putting $y = e^{mx}$, we have, $Dy = me^{mx}, \cdots, D^r y = m^r e^{mx}$;

hence, $$f(D)(e^{mx}) = e^{mx} f(m).$$

For e^{mx} to be an integral of (2), m must satisfy the equation

(3) $$f(m) = 0,$$

i.e. $$k_0 m^n + k_1 m^{n-1} + k_2 m^{n-2} + \cdots + k_{n-1} m + k_n = 0.$$

Each value of m satisfying (3) gives an integral of (2). If these are all distinct (say $m_1, m_2, m_3, \cdots, m_n$), $e^{m_1 x}, e^{m_2 x}, \cdots, e^{m_n x}$ will be linearly independent, and making use of *A*, § 42, $c_1 e^{m_1 x} + c_2 e^{m_2 x} + \cdots + c_n e^{m_n x}$ will be the general integral of (2), and the complementary function of (1).

Remark. — Equation (3), which is so readily obtained from equation (2), is usually referred to as the *auxiliary equation.**

Ex. 1. $\dfrac{d^2y}{dx^2} - 3\dfrac{dy}{dx} + 2y = 0.$

The auxiliary equation is $m^2 - 3m + 2 = 0$. Its roots are 1, 2. Hence the general solution is

$$y = c_1 e^x + c_2 e^{2x}.$$

Ex. 2. $\dfrac{d^2y}{dx^2} - 6\dfrac{dy}{dx} + 25y = 0.$

Here $m^2 - 6m + 25 = 0$, whence $m = 3 \pm 4i$, where $i = \sqrt{-1}$.

$$\therefore y = c_1 e^{(3+4i)x} + c_2 e^{(3-4i)x}, \quad \text{or } y = e^{3x}(c_1 e^{4ix} + c_2 e^{-4ix}).$$

Ex. 3. $\dfrac{d^3y}{dx^3} - \dfrac{dy}{dx} = 0.$

Ex. 4. $(D^3 - 2D^2 - D + 2)y = 0.$

* Cauchy calls this the *characteristic equation.*

44. Roots of Auxiliary Equation Repeated. — If any roots of the auxiliary equation are repeated, the method of § 43 does not give us n linearly independent integrals, and consequently it does not give us the general solution. In this case we make the more general substitution $y = e^{mx}\phi(x)$, where $\phi(x)$ is a function of x entirely at our disposal. Then

$$Dy = e^{mx}[m\phi + D\phi],$$

$$D^2y = e^{mx}[m^2\phi + 2\,mD\phi + D^2\phi],$$

$$D^3y = e^{mx}[m^3\phi + 3\,m^2D\phi + 3\,mD^2\phi + D^3\phi],$$

$$\cdot \quad \cdot \quad \cdot \quad \cdot \quad \cdot \quad \cdot \quad \cdot \quad \cdot \quad \cdot \quad \cdot \quad \cdot$$

$$D^ny = e^{mx}\Big[m^n\phi + nm^{n-1}D\phi + \frac{n(n-1)}{2\,!}m^{n-2}D^2\phi + \cdots$$
$$+ \frac{n(n-1)\cdots(n-r+1)}{r\,!}m^{n-r}D^r\phi + \cdots + D^n\phi\Big],$$

whence

$$f(D)\,y = e^{mx}\Big[f(m)\phi + f'(m)D\phi + \frac{1}{2\,!}f''(m)D^2\phi + \cdots$$
$$+ \frac{1}{r\,!}f^{(r)}(m)\,D^r\phi + \cdots + \frac{1}{n\,!}f^{(n)}(m)\,D^n\phi\Big]$$

where $f'(m) = \dfrac{d}{dm}f(m), \cdots, f^{(r)}(m) = \dfrac{d^r}{dm^r}f(m)$.

If m_1 is an r-fold root of $f(m) = 0$, we have § (31),

$$f(m_1) = 0,\ f'(m_1) = 0,\ \cdots,\ f^{(r-1)}(m) = 0.$$

In this case $f(D)\,y$ will vanish if $y = e^{mx}\phi(x)$ provided $D^r\phi = 0$, whence all the higher derivatives of ϕ are also zero; *i.e.* provided $\phi = c_1x^{r-1} + c_2x^{r-2} + \cdots + c_{r-1}x + c_r$, where $c_1, c_2, \cdots, c_r$ are any constants whatever. Hence, we see that *if* m_1 *corresponds to an* *r-fold root of the auxiliary equation, not only is e^{m_1x} an integral of the equation, but so also are xe^{m_1x}, $x^2e^{m_1x}$, $\cdots$, $x^{r-1}e^{m_1x}$, i.e.* corresponding to an r-fold root we have r linearly independent integrals. So that whether the roots of the auxiliary equation are repeated or not, the n linearly independent integrals necessary for obtaining the complementary function (A, § 42) are always supplied by the auxiliary equation.

Ex. 1. $(4D^3 - 3D + 1)y = 0$.

The roots of $4m^3 - 3m + 1 = 0$ are $\frac{1}{2}, \frac{1}{2}, -1$. Hence the general solution is $y = e^{\frac{1}{2}x}(c_1 + c_2x) + c_3e^{-x}$.

Ex. 2. $(D^3 - D^2 - D + 1)y = 0$.

Ex. 3. $(D^4 + 2D^3 - 2D - 1)y = 0$.

Ex. 4. $(D^3 - 6D^2 + 9D)y = 0$.

45. Roots of the Auxiliary Equation Complex. If the coefficients of the differential equation are real, while some or all of the roots of auxiliary equation are not, we can, by a proper arrangement of the terms in the complementary function, have the latter involve only real terms. Thus, if the auxiliary equation has a root $\alpha + i\beta$, it will also have $\alpha - i\beta$ as a root, since its coefficients are real. Two terms of the complementary function will then be

$$c_1e^{(\alpha+i\beta)x} + c_2e^{(\alpha-i\beta)x},$$

or

$$e^{\alpha x}(c_1e^{i\beta x} + c_2e^{-i\beta x}).$$

Now $e^{i\beta x} = \cos\beta x + i\sin\beta x$, and $e^{-i\beta x} = \cos\beta x - i\sin\beta x$. Hence our pair of terms may be written

$$e^{\alpha x}[(c_1 + c_2)\cos\beta x + i(c_1 - c_2)\sin\beta x].$$

Putting $c_1 = \dfrac{A - iB}{2}$, $c_2 = \dfrac{A + iB}{2}$, this becomes

$$e^{\alpha x}(A \cos \beta x + B \sin \beta x),$$

where A and B are the two arbitrary constants.

Another form in which this may be written is $ae^{\alpha x} \sin(\beta x + b)$,* or $ae^{\alpha x} \cos(\beta x + b)$, where a and b are the arbitrary constants. For interpreting the solutions in physical problems, the latter forms are sometimes preferable.

It is obvious, that in case a pair of such roots is repeated, the corresponding part of the complementary function is

$$e^{\alpha x}(A_1 \cos \beta x + B_1 \sin \beta x) + xe^{\alpha x}(A_2 \cos \beta x + B_2 \sin \beta x)$$

or $$e^{\alpha x}[(A_1 + A_2 x) \cos \beta x + (B_1 + B_2 x) \sin \beta x].$$

And perfectly generally, in case such a pair occur as r-fold roots, the corresponding part of the complementary function is

$$e^{\alpha x}[(A_1 + A_2 x + \cdots + A_r x^{r-1}) \cos \beta x + (B_1 + B_2 x + \cdots + B_r x^{r-1}) \sin \beta x].$$

Ex. 1. In the case of Ex. 2, § 43, $\alpha = 3$, $\beta = 4$, so that the solution may also be written

$$y = e^{3x}(A \cos 4x + B \sin 4x)$$

or $$y = a e^{3x} \cos(4x + b).$$

Ex. 2. $(D^4 + 2D^2 + 1)y = 0.$

Ex. 3. $(D^3 - D^2 + D)y = 0.$

*For, $A \cos \beta x + B \sin \beta x$ may be written $\sqrt{A^2 + B^2}\left(\dfrac{A}{\sqrt{A^2+B^2}} \cos \beta x + \dfrac{B}{\sqrt{A^2+B^2}} \sin \beta x\right)$. Since the sum of the squares of $\dfrac{A}{\sqrt{A^2+B^2}}$ and $\dfrac{B}{\sqrt{A^2+B^2}}$ equals unity, these may be taken as the sine and cosine of some angle, say b. Putting $\sqrt{A^2 + B^2} = a$, our expression becomes $a(\sin b \cos \beta x + \cos b \sin \beta x)$ or $a \sin(\beta x + b)$.

Remark.—For the purpose of interpreting the solution of certain problems in Physics it is desirable at times to introduce hyperbolic functions in place of the exponentials in case a pair of the roots of the auxiliary equation are real and equal to within the sign. Proceeding as before, we make use of the formulæ $e^x = \cosh x + \sinh x$, $e^{-x} = \cosh x - \sinh x$.

If $+m$ and $-m$ are a pair of roots of the auxiliary equation, the corresponding terms of the complementary function are

$$\begin{aligned} y &= c_1 e^{mx} + c_2 e^{-mx} \\ &= (c_1 + c_2)\cosh mx + (c_1 - c_2)\sinh mx \\ &= A\cosh mx + B\sinh mx. \end{aligned}$$

Using the addition theorem for the hyperbolic functions, this may also be written

$$y = a\cosh(mx + b),$$

or

$$y = a\sinh(mx + b),$$

where a and b are arbitrary constants.

46. Properties of the Symbolic Operator $(D - \alpha)$.—

1° $(D-\alpha)y$ means $\dfrac{dy}{dx} - \alpha y$. Similarly $(D-\beta)y$ means $\dfrac{dy}{dx} - \beta y$. Hence $[(D-\alpha)+(D-\beta)]y$ means $2\dfrac{dy}{dx} - (\alpha+\beta)y$, which may be written symbolically $[2D - (\alpha+\beta)]y$. That is, the result of operating on y with $(D-\alpha)$ and $(D-\beta)$ separately and then taking the sum, is the same as operating on y with $[2D - (\alpha+\beta)]$. Hence we see that *the operation resulting from taking the sum of the results of two operations of the type here considered can be gotten symbolically by taking the sum of their symbolic representatives.* Thus we can write

$$[(D-\alpha)+(D-\beta)] = [2D - (\alpha+\beta)].$$

Evidently this rule applies to the sum of any number of such operators, and also to the difference between any two of them.

2° $(D-\beta)(D-\alpha)y$ means $\left(\dfrac{d}{dx} - \beta\right)\left(\dfrac{dy}{dx} - \alpha y\right)$, which is $\dfrac{d^2y}{dx^2} - (\alpha+\beta)\dfrac{dy}{dx} + \alpha\beta y$. That is, the result of operating on y with

$(D-\alpha)$ first, and then with $(D-\beta)$ on the result, is the same as operating on y with $[D^2-(\alpha+\beta)D+\alpha\beta]$. Hence we see that *the operation resulting from the successive performance of two operations of the type here considered, can be gotten symbolically by taking the product of their symbolic representatives.* Thus we can write

$$[(D-\beta)(D-\alpha)]=[D^2-(\alpha+\beta)D+\alpha\beta].$$

Moreover, owing to the symmetry of α and β in the right-hand member, we see that *the order of the operators on the left is not essential,* or, as it is usually put, *two operations of the type here considered are commutative.*

Obviously all this applies to any number of operations of the type here considered.

All the results of this paragraph can be incorporated in the following:

The symbolic representatives of operations of the type here considered behave like algebraic quantities for the processes of addition, subtraction, and multiplication.

Remark.—Since any polynomial in D with constant coefficients is a product of linear factors, this theorem applies also to operators whose symbolic representatives are polynomials in D with constant coefficients.

Evidently if the roots of the auxiliary equation of (1) are m_1, m_2, $\cdots$, m_n (whether these are all distinct or not), we may write (1) in the form

$$k_0(D-m_1)(D-m_2)\cdots(D-m_n)y=X.$$

47. Particular Integral.—A perfectly general method for obtaining the particular integral of a complete linear differential equation with constant coefficients (and, for that matter, another method for obtaining the complementary function, as well) results from the following considerations:

In the following discussion we shall suppose the equation divided through by k_0, and to simplify matters and yet bring out the method, we shall use an equation of the third order. Let us start, then, with the equation

$$f(D)y = (D - m_1)(D - m_2)(D - m_3)y = X.$$

To find the integral of this equation is to find y, a function of x, such that when operated on by $f(D)$ it will give X.

Let $(D - m_2)(D - m_3)y = u$, where u is a new function. Then

$$(D - m_1)u = X, \text{ or } \frac{du}{dx} - m_1 u = X.$$

This is a linear equation of the first order, and e^{-m_1x} is an integrating factor (§ 13). Hence

$$e^{-m_1x}u = \int e^{-m_1x}X\,dx + c, \text{ or } u = e^{m_1x}\int e^{-m_1x}X\,dx + ce^{m_1x};$$

i.e.
$$(D - m_2)(D - m_3)y = e^{m_1x}\int e^{-m_1x}X\,dx + ce^{m_1x}.$$

Now let
$$(D - m_3)y = v.$$

Then
$$(D - m_2)v = e^{m_1x}\int e^{-m_1x}X\,dx + ce^{m_1x}.$$

This is also linear and of the first order, hence an integrating factor is e^{-m_2x}. Introducing this, we have

$$ve^{-m_2x} = \int e^{(m_1-m_2)x}\left[\int e^{-m_1x}X\,dx\right]dx + \frac{c}{m_1 - m_2}e^{(m_1-m_2)x} + c',$$

or
$$v = e^{m_2x}\int e^{(m_1-m_2)x}\left[\int e^{-m_1x}X\,dx\right]dx + \frac{c}{m_1 - m_2}e^{m_1x} + c'e^{m_2x}.$$

Hence $(D-m_3)y = e^{m_2x}\int e^{(m_1-m_2)x}\left[\int e^{-m_1x}X\,dx\right]dx + c''e^{m_1x} + c'e^{m_2x},$

where $$c'' = \frac{c'}{m_1 - m_2}.$$

This is again linear, and an integrating factor is e^{-m_3x}. Using this we have

$$ye^{-m_3x} = \int e^{(m_2-m_3)x}\left\{\int e^{(m_1-m_2)x}\left[\int e^{-m_1x}X\,dx\right]dx\right\}dx$$

$$+\frac{c''}{m_1-m_3}e^{(m_1-m_3)x} + \frac{c'}{m_2-m_3}e^{(m_2-m_3)x} + c_3,$$

or $$y = e^{m_3x}\int e^{(m_2-m_3)x}\left\{\int e^{(m_1-m_2)x}\left[\int e^{-m_1x}X\,dx\right]dx\right\}dx$$

$$+ c_1e^{m_1x} + c_2e^{m_2x} + c_3e^{m_3x}.$$

This law in the case of the nth order is obvious now.* It is

I. $$y = e^{m_nx}\int e^{(m_{n-1}-m_n)x}\int\cdots\int e^{(m_1-m_2)x}\int e^{-m_1x}X(dx)^n$$

$$+ c_1e^{m_1x} + c_2e^{m_2x} + \cdots + c_ne^{m_nx}.$$

Remark. — In the second line we have the complementary function, with which we are already familiar (§ 43). (Let the student show that in the case of repeated roots of the auxiliary equation this method leads to the same result as § 44.) In the first line we have the particular integral, whether the roots of the auxiliary equation are all distinct or not.

* To prove this, we simply need assume it for the nth order, and show that it holds for the $(n+1)$st order. This can be done at once, and will be left as an exercise to the student.

Ex. 1. $\dfrac{d^3y}{dx^3} - \dfrac{d^2y}{dx^2} - 2\dfrac{dy}{dx} = e^{-x}.$

The auxiliary equation is

$$m^3 - m^2 - 2m = 0. \quad \therefore m = 0, \ -1, \ 2.$$

The complementary function is $Y = c_1 + c_2e^{-x} + c_3e^{2x}$.

The particular integral is

$$U = e^{2x}\int e^{(-1-2)x}\int e^{(0+1)x}\int e^{-x}(dx)^3$$

$$= e^{2x}\int e^{-3x}\int e^{x}(-e^{-x})(dx)^2 = -e^{2x}\int e^{-3x}\left[\int dx\right]dx$$

$$= -e^{2x}\int e^{-3x}x\,dx = \frac{1}{3}xe^{-x} + \frac{1}{9}e^{-x}.$$

Since e^{-x} is already part of the complementary function, it will be sufficient to use $\frac{1}{3}xe^{-x}$, thus giving the general solution,

$$y = c_1 + c_2e^{-x} + c_3e^{2x} + \frac{1}{3}xe^{-x}.$$

Ex. 2. $(D^2 + 3D + 2)y = e^{e^x}$.

Ex. 3. $(D^3 + 3D^2 + 3D + 1)y = 2e^{-x} - x^2e^{-x}$.

Ex. 4. $(D^2 - D - 2)y = \sin x$.

Ex. 5. $(D - 1)^2y = \dfrac{e^x}{(1 - x)^2}$.

48. Another Method of finding the Particular Integral.* — The general method of finding the particular integral given in § 47 is frequently long. At times, the first integration is readily obtained, but the successive ones are long and tedious. In such cases the following method applies:

Starting with $$(D-m_1)(D-m_2)\cdots(D-m_n)y=X,$$

we can write symbolically

$$y=\frac{1}{(D-m_1)(D-m_2)\cdots(D-m_n)}X,$$

where $\frac{1}{(D-m_1)(D-m_2)\cdots(D-m_n)}$ is the symbol of the operation inverse to $(D-m_1)(D-m_2)\cdots(D-m_n)$. Just as $\sin^{-1}x$ means such a function of x that when we operate on it with the operator sin we get x, so if we operate on

$$\frac{1}{(D-m_1)(D-m_2)\cdots(D-m_n)}X$$

with $(D-m_1)(D-m_2)\cdots(D-m_n)$, we get X. Now we have seen that the operator $(D-m_1)(D-m_2)\cdots(D-m_n)$ is equivalent to the successive performance of the operators $(D-m_1)$, $(D-m_2)$, $\cdots$, $(D-m_n)$; and besides, the order of the latter is not essential.

Looked upon algebraically the fraction

$$\frac{1}{(D-m_1)(D-m_2)\cdots(D-m_n)}$$

is equal to the sum of the partial fractions

$$\frac{a_1}{D-m_1}+\frac{a_2}{D-m_2}+\cdots+\frac{a_n}{D-m_n},$$

if the roots of the auxiliary equation are distinct.

* This was first published by Lobatto, *Théorie des Caractéristiques*, Amsterdam, 1837. Independently it was given by *Boole*, *Cambridge Math. Journal*, 1st series, Vol. II, p. 114.

Looked upon as operators, this equality still holds; for to verify the equality we operate on both with $(D-m_1)(D-m_2)\cdots(D-m_n)$. Since the order in which we operate with these factors is immaterial, the result will be that all the operators resulting are polynomials, which can be treated as algebraic expressions. Hence the algebraic equality of the symbolic representatives of the two operators means the equality of the operators, and the original operators, in fractional form, are also equivalent; *i.e.*

$$y=\frac{1}{(D-m_1)(D-m_2)\cdots(D-m_n)}X=\frac{a_1}{D-m_1}X+\frac{a_2}{D-m_2}X+\cdots$$
$$+\frac{a_n}{D-m_n}X.$$

If we put $u=\frac{a}{D-m}X$, then $(D-m)u=aX$.

Integrating this linear equation, we have $ue^{-mx}=a\int e^{-mx}X\,dx$, or $u=ae^{mx}\int e^{-mx}X\,dx$. Hence the particular integral may be put in the form

$$\text{II.}\quad a_1e^{m_1x}\int e^{-m_1x}X\,dx+a_2e^{m_2x}\int e^{-m_2x}X\,dx+\cdots+a_ne^{m_nx}\int e^{-m_nx}X\,dx.$$

Remark 1.—This method leads to a real particular integral, even in case a pair of the roots of the auxiliary equation are conjugate complex quantities, $\alpha+i\beta$ and $\alpha-i\beta$. In breaking up $\frac{1}{f(D)}$ into a sum of partial fractions, we know that the sum $\frac{a_1}{D-(\alpha+i\beta)}+\frac{a_2}{D-(\alpha-i\beta)}$ is equal to $\frac{kD+l}{(D-\alpha)^2+\beta^2}$, in which k and l are real. Hence a_1 and a_2 are also conjugate complex quantities, say $\lambda+i\mu$ and $\lambda-i\mu$.

$$\text{Now}\ \frac{\lambda+i\mu}{D-(\alpha+i\beta)}X=(\lambda+i\mu)e^{(\alpha+i\beta)x}\int e^{-(\alpha+i\beta)x}X\,dx$$
$$=(\lambda+i\mu)e^{\alpha x}(\cos\beta x+i\sin\beta x)\int e^{-\alpha x}X(\cos\beta x-i\sin\beta x)dx.$$

Since $\frac{\lambda - i\mu}{D-(\alpha - i\beta)}X$ may be gotten from the above by changing the sign of i wherever it occurs, the two have the same real parts, while their imaginary parts are equal but of opposite signs. Hence their sum is equal to twice the real part of either; *i.e.*

$$\frac{\lambda + i\mu}{D-(\alpha+i\beta)}X + \frac{\lambda - i\mu}{D-(\alpha - i\beta)}X = 2\, e^{\alpha x}(\lambda \cos\beta x - \mu \sin\beta x)\int e^{-\alpha x} X \cos\beta x\, dx$$

$$+ 2\, e^{\alpha x}(\lambda \sin\beta x + \mu \cos\beta x)\int e^{-\alpha x} X \sin\beta x\, dx.$$

Remark 2.—In case a root is repeated the following obvious modification is necessary:

To fix the ideas we shall suppose one of the roots, m_1, is a triple root. The corresponding partial fractions will be $\frac{a_1}{D-m_1} + \frac{a_2}{(D-m_1)^2} + \frac{a_3}{(D-m_1)^3}$, and the corresponding terms of the particular integral will be (I, § 47)

$$a_1 e^{m_1 x}\int e^{-m_1 x} X\, dx + a_2 e^{m_1 x}\int\int e^{-m_1 x} X\, (dx)^2 + a_3 e^{m_1 x}\int\int\int e^{-m_1 x} X\, (dx)^3.$$

Ex. 1. $(D^2 - 3D + 2)y = e^x.$

Ex. 2. $(D^3 - 3D^2 - D + 3)y = x^2.$

Ex. 3. $(D^2 + 1)y = \sec x.$

Ex. 4. $(D^3 - 4D^2 + 5D - 2)y = x.$

49. Variation of Parameters.* — Another general method of obtaining the particular integral, known as the method of *variation of parameters*, at times applies very readily, especially if the order of the equation is not high. The method consists in considering the constants in the complementary function no longer as constants, but as undetermined functions of x such that when substituted in $f(D)y$ we get X, and not zero, as is the case when they are constants.

* This method is due to Joseph Louis Lagrange (1736–1813).

Since we have n functions at our disposal, and only one condition to impose upon them, it is clear that, theoretically at least, we can satisfy this requirement in an indefinite number of ways, by imposing any other $n-1$ conditions we please. In actual practice we shall impose these conditions in such a way as to simplify our work as much as possible.

The method will be carried out in the case of an equation of the third order. (The argument will readily be seen to apply to any order.)

Let the equation be

$$(1) \qquad (k_0 D^3 + k_1 D^2 + k_2 D + k_3)y = X,$$

and let the complementary function be

$$(2) \qquad y = c_1 e^{m_1 x} + c_2 e^{m_2 x} + c_3 e^{m_3 x}.*$$

We shall try to find c_1, c_2, c_3, such that (2) shall be a solution of (1). This still allows us to impose two conditions upon c_1, c_2, c_3.

Differentiating (2), we get

$$Dy = m_1 c_1 e^{m_1 x} + m_2 c_2 e^{m_2 x} + m_3 c_3 e^{m_3 x} + e^{m_1 x}\frac{dc_1}{dx} + e^{m_2 x}\frac{dc_2}{dx} + e^{m_3 x}\frac{dc_3}{dx}.$$

We shall now use one of the two conditions at our disposal by letting

$$(3) \qquad e^{m_1 x}\frac{dc_1}{dx} + e^{m_2 x}\frac{dc_2}{dx} + e^{m_3 x}\frac{dc_3}{dx} = 0;$$

so that we have

$$(4) \qquad Dy = m_1 c_1 e^{m_1 x} + m_2 c_2 e^{m_2 x} + m_3 c_3 e^{m_3 x}.$$

* If any of the roots of the auxiliary equation are repeated or imaginary, the resulting change in the form of the complementary function causes no difference in the process.

Differentiating, we get

$$D^2y = m_1^2c_1e^{m_1x} + m_2^2c_2e^{m_2x} + m_3^2c_3e^{m_3x} + m_1e^{m_1x}\frac{dc_1}{dx} + m_2e^{m_2x}\frac{dc_2}{dx} + m_3e^{m_3x}\frac{dc_3}{dx}.$$

Here again we put

$$(5) \qquad m_1e^{m_1x}\frac{dc_1}{dx} + m_2e^{m_2x}\frac{dc_2}{dx} + m_3e^{m_3x}\frac{dc_3}{dx} = 0,$$

thus using the second condition still at our disposal, so that

$$(6) \qquad D^2y = m_1^2c_1e^{m_1x} + m_2^2c_2e^{m_2x} + m_3^2c_3e^{m_3x}.$$

Differentiating again, we get

$$(7) \qquad D^3y = m_1^3c_1e^{m_1x} + m_2^3c_2e^{m_2x} + m_3^3c_3e^{m_3x} + m_1^2e^{m_1x}\frac{dc_1}{dx} + m_2^2e^{m_2x}\frac{dc_2}{dx} + m_3^2e^{m_3x}\frac{dc_3}{dx}.$$

Substituting (2), (4), (6), (7), in (1), and remembering that with c_1, c_2, c_3 constant, (2) is the complementary function, we have

$$(8) \qquad k_0m_1^2e^{m_1x}\frac{dc_1}{dx} + k_0m_2^2e^{m_2x}\frac{dc_2}{dx} + k_0m_3^2e^{m_3x}\frac{dc_3}{dx} = X.$$

Equations (3), (5), (8) are three linear equations sufficient to determine $\frac{dc_1}{dx}$, $\frac{dc_2}{dx}$, $\frac{dc_3}{dx}$, and by quadratures c_1, c_2, c_3 will be found such that (2) will be a solution of (1), the constants of integration giving us again the complementary function.

The method of variation of parameters applies to all linear equations, whether the coefficients are constants or not. (Thus, see Ex. 4, § 53.) As an illustration, we shall solve the general linear differential equation of the first order (§ 13) by this method.

The general equation is

(1) $$\frac{dy}{dx} + Py = Q,$$

where P and Q are functions of x. Let us consider, first,

(2) $$\frac{dy}{dx} + Py = 0, \text{ or}$$

$$\frac{dy}{y} + P\,dx = 0.$$

Integrating, we get $\log y + \int P dx = C$, or

(3) $$ye^{\int Pdx} = e^C = c.$$

Now let c be considered a function of x. We shall determine it, so that (3) shall satisfy (1). Differentiating, we have

$$e^{\int Pdx}\left(\frac{dy}{dx} + Py\right) = \frac{dc}{dx}.$$

Comparing this with (1), we see that we must have

$$\frac{dc}{dx} = Qe^{\int Pdx}, \text{ or } c = \int Qe^{\int Pdx}\,dx + c'.$$

$\therefore ye^{\int Pdx} = \int Qe^{\int Pdx}\,dx + c'$ is the solution (the result we obtained in § 13).

Ex. 1. $\frac{d^2y}{dx^2} + y = \sec x$. (Ex. 3, § 48.)

The roots of the auxiliary equation are $\pm i$. Hence the complementary function is

$$y = c_1 \cos x + c_2 \sin x.$$

$$Dy = -c_1 \sin x + c_2 \cos x + \left| \cos x \frac{dc_1}{dx} + \sin x \frac{dc_2}{dx} = 0.\right.$$

$$D^2y = -c_1 \cos x - c_2 \sin x - \sin x \frac{dc_1}{dx} + \cos x \frac{dc_2}{dx}.$$

Substituting in the differential equation, we have

$$-\sin x \frac{dc_1}{dx} + \cos x \frac{dc_2}{dx} = \sec x.$$

Besides, we chose $\cos x \frac{dc_1}{dx} + \sin x \frac{dc_2}{dx} = 0.$

$$\therefore \frac{dc_1}{dx} = -\sin x \sec x = -\frac{\sin x}{\cos x}, \qquad c_1 = \log \cos x + C_1.$$

$$\frac{dc_2}{dx} = 1, \qquad c_2 = x + C_2.$$

And the complete solution is

$$y = C_1 \cos x + C_2 \sin x + \cos x \log \cos x + x \sin x.$$

Ex. 2. $\frac{d^2y}{dx^2} + y = \tan x.$

50. Method of Undetermined Coefficients. — We shall conclude the discussion of the problem of finding the particular integral with an account of a method, which, while not applicable in all cases, is relatively simple whenever it can be used. It applies to all cases in which the right-hand member contains only terms which have a finite number of distinct derivatives. Such terms are x^h, e^{lx}, $\sin mx$, $\cos nx$, and products of these, where h is any positive integer, and l, m, n are any constants.

By this method we find the particular integral U by inspection, or by trial, as it were.

If we take, as a first trial, the terms of the right-hand member X, each prefixed by an undetermined multiplier, we shall find that, as a rule, on substituting this in the left-hand member, $f(D)y$, other terms arise as a result of differentiation. Consequently, we shall use for U the sum of all the terms of X, together with all those arising from them by differentiation (by hypothesis there are only a finite number of these), each prefixed by an undetermined multiplier. We then equate identically to X the result of this substitution (*i.e.* we equate the coefficients of corresponding terms). This will give as many equations of condition among the undetermined multipliers as there are distinct terms in $f(D)\,U$. This number is either equal to or less than the number of undetermined multipliers (for all the terms obtainable from U by differentiation need not occur), and these multipliers can then be calculated.

Ex. 1. $\dfrac{d^2y}{dx^2}+4y=x^2+\cos x.$

The roots of the auxiliary equation are $\pm 2i$.

$$\therefore\ Y=A\cos 2x+B\sin 2x.$$

For a particular integral we take,

$$U=ax^2+bx+c+f\cos x+g\sin x.$$

Then, $$D^2U=\qquad 2a-f\cos x-g\sin x.$$

$$\therefore f(D)\,U=4ax^2+4bx+2a+4c+3f\cos x+3g\sin x.$$

Equating coefficients of this to those of X we have,

$$4a=1 \qquad \therefore\ a=\frac{1}{4},$$

$$4b=0 \qquad b=0,$$

$$2a+4c=0 \qquad c=-\frac{1}{8},$$

$$3f=1 \qquad f=\frac{1}{3},$$

$$3g=0 \qquad g=0.$$

Hence the general solution is

$$y = A\cos 2x + B\sin 2x + \frac{1}{4}x^2 - \frac{1}{8} + \frac{1}{3}\cos x.$$

Ex. 2. $(D^2 - 2D + 1)y = 2xe^{2x} - \sin^2 x.$

The complementary function is readily seen to be

$$Y = (c_1 + c_2x)e^x.$$

To find the particular integral it will be simpler to replace $-\sin^2 x$ by $\frac{1}{2}\cos 2x - \frac{1}{2}$. Doing this, we put

$$U = axe^{2x} + be^{2x} + c\cos 2x + f\sin 2x + g.$$

$$\therefore\ DU = 2axe^{2x} + (a + 2b)e^{2x} + 2f\cos 2x - 2c\sin 2x.$$

$$D^2U = 4axe^{2x} + 4(a + b)e^{2x} - 4c\cos 2x - 4f\sin 2x.$$

$$f(D)U = axe^{2x} + (2a + b)e^{2x} - (3c + 4f)\cos 2x - (3f - 4c)\sin 2x + g$$

Hence we must have

$$a = 2 \qquad \therefore\ a = 2,$$

$$2a + b = 0 \qquad b = -4,$$

$$3c + 4f = -\frac{1}{2} \qquad c = -\frac{3}{50},$$

$$3f - 4c = 0 \qquad f = -\frac{2}{25},$$

$$g = -\frac{1}{2} \qquad g = -\frac{1}{2},$$

and the general solution is

$$y = (c_1 + c_2x)e^x + 2xe^{2x} - 4e^{2x} - \frac{3}{50}\cos 2x - \frac{2}{25}\sin 2x - \frac{1}{2}.$$

Ex. 3. $(D^2+1)y = 2e^x + x^3 - x.$

Ex. 4. $(D^2+2D+1)y = 3e^{2x} - \cos x.$

Ex. 5. $(D^3-1)y = x^2.$

This method will be in default in either of the following two cases:

1° If a term in the right-hand member is also a term in the complementary function, it is clear that the substitution of such a term or any of its derivatives for y in $f(D)y$ will not give rise to that term. As a matter of fact we get zero.

We shall first suppose that the root of the auxiliary equation, to which the term in question corresponds, is a simple one; if u is the term, this amounts to having

$$f(D)u = 0, \text{ but } f'(D)u \neq 0,$$

where $f'(D)$ is $\dfrac{df(D)}{dD}$.*

Now since $f(D)$ is a polynomial in D, and since $D^k(xu) = xD^ku + kD^{k-1}u$ it is clear that $f(D)(xu) = xf(D)u + f'(D)u$.

Since $f'(D)u$, by hypothesis, is different from zero, it follows that if for y in $f(D)y$ we substitute xu + terms derived from this by differentiation, we shall have resulting the term u + terms arising from it by differentiation, and none other.

Perfectly generally, if the term u is also a term in the complementary function which corresponds to an r-fold root of the auxiliary equation, then

$$f(D)u = 0, f'(D)u = 0, \cdots, f^{(r-1)}(D)u = 0, \text{ but } f^{(r)}(D)u \neq 0.$$

* Letting m be the root corresponding to u, we have (§ 31) that $f'(m) \neq 0$ if m is a simple root of $f(m) = 0$. Hence u is not an integral of the equation $f'(D)y = 0$.

Since

$$D^k(x^r u) = x^r D^k u + krx^{r-1}D^{k-1}u + \frac{k(k-1)}{2}r(r-1)x^{r-2}D^{k-2}u +$$
$$\cdots + \frac{k(k-1)\cdots(k-s+1)}{s\,!}r(r-1)\cdots(r-s+1)x^{r-s}D^{k-s}u +$$
$$\cdots + k(k-1)\cdots(k-r+1)D^{k-r}u,$$

and since $f(D)$ is a polynomial in D with constant coefficients, it follows that

$$f(D)(x^r u) = x^r f(D)u + rx^{r-1}f'(D)u + \frac{r(r-1)}{2}x^{r-2}f''(D)u + \cdots$$
$$+ \frac{r(r-1)\cdots 3\cdot 2}{(r-1)\,!}xf^{(r-1)}(D)u + f^{(r)}(D)u.$$

All of the terms on the right are zero, except $f^{(r)}(D)u$, which is definitely not zero. Hence if for y in $f(D)y$ we substitute $x^r u +$ terms arising from this by differentiation, we shall obtain the term $u +$ terms arising from it by differentiation, and none other.

2° The second case where the original method will be at fault is where terms of the type $x^t u$ occur, u being a term in the complementary function. A similar modification of the method applies here. Suppose that u corresponds to an r-fold root of the auxiliary equation.* As before, we have

$$f(D)(x^{t+r}u) = x^{t+r}f(D)u + (t+r)x^{t+r-1}f'(D)u$$
$$+ \frac{(t+r)(t+r-1)}{2\,!}x^{t+r-2}f''(D)u + \cdots$$
$$+ \frac{(t+r)(t+r-1)\cdots(t+2)}{(r-1)\,!}x^{t+1}f^{(r-1)}(D)u$$
$$+ \frac{(t+r)(t+r-1)\cdots(t+1)}{r\,!}x^t f^{(r)}D(u).$$

* Of course xu, x^2u, $\cdots$, $x^{r-1}u$ are also terms in the complementary function. In this discussion, u is supposed to be that term which does not contain x as a factor otherwise the exponent t would be indeterminate.

All of these terms on the right are zero, except $f^{(r)}(D)u$, which is definitely not zero. Hence, if for y in $f(D)y$ we substitute $x^{t+r}u +$ terms arising from it by differentiation, we shall obtain $x^t u +$ terms arising from it by differentiation, and none other.

We are now in a position to formulate our rule:

When the right-hand member of the differential equation contains only terms which have a finite number of distinct derivatives, take for particular integral the sum of all the terms together with all those obtained from these by differentiation, prefixing to each of them an undetermined coefficient. These coefficients are determined by substituting the trial particular integral in the differential equation and equating coefficients. In case a term in the right-hand member is a term in the complementary function or a term in it multiplied by an integral power of x, which term corresponds to an r-fold root of the auxiliary equation, replace that term in the right-hand member by x^r times it in making up the trial particular integral.

Remark.—It may not always be necessary to insert all the terms suggested by the general rule. These can frequently be detected by inspection. Thus in Ex. 1, since the coefficient of $\frac{dy}{dx}$ in the differential equation is zero, the terms x and $\sin x$ in the trial particular integral are unnecessary, for these will obviously not appear as a result of substituting $ax^2 + f \cos x$ in the equation. If any unnecessary terms are put in the trial particular integral, that fact will show itself by having their coefficients turn out to be zero. So that, excepting the unnecessary labor, the introduction of extraneous terms in the trial integral is not serious. It is also obviously useless to put in any terms which appear in the complementary function. (If such a term is included in the trial particular integral its coefficient will, of course, not appear in the resulting equations among the coefficients. This means that this coefficient may be chosen arbitrarily; which is exactly as it should be.) As a consequence, when any term in the right-hand member is replaced by x^r times it in either of the two exceptional cases referred to in the rule, only those terms obtained from this by differentiation which are not in the complementary function need be added.

Ex. 6. $(D^3 - 2D^2 - 3D)y = 3x^2 + \sin x$.

Hint. — Since 0 is a simple root of the auxiliary equation, and 1 is therefore a term in the complementary function, we shall have to try $ax^3 + bx^2 + cx$ to get x^2. Moreover, $\sin x$ is not a part of the complementary function. Hence the trial integral is $U = ax^3 + bx^2 + cx + f \sin x + g \cos x$.

Ex. 7. $(D^4 - 2D^2 + 1)y = e^x + 4$.

Ex. 8. $(D^2 - 2D)y = e^{2x} + 1$.

Ex. 9. $(D^4 + 2D^2 + 1)y = \cos x$.

51. Cauchy's Linear Equation. — The linear differential equation

$$(1) \qquad k_0 x^n \frac{d^n y}{dx^n} + k_1 x^{n-1} \frac{d^{n-1} y}{dx^{n-1}} + \cdots + k_{n-1} x \frac{dy}{dx} + k_n y = X,^*$$

where the coefficient of $\frac{d^r y}{dx^r}$ is a constant times x^r, is at once reducible by the transformation $x = e^z$ to an equation with constant coefficients. For

$$\frac{dy}{dx} = \frac{dy}{dz}\frac{dz}{dx} = \frac{1}{x}\frac{dy}{dz},$$

$$\frac{d^2y}{dx^2} = \qquad = \frac{1}{x^2}\left(\frac{d^2y}{dz^2} - \frac{dy}{dz}\right),$$

* This form of linear equation is often called the homogeneous linear equation. This seems rather unfortunate. I prefer to reserve this name for any linear equation which is homogeneous in y and its derivatives, in conformity with a large number of writers on the subject, and shall refer to the above linear equation (1) as the *Cauchy linear differential equation*, after Augustin Louis Cauchy (1789-1857). See his *Exercises d'Analyse.*

$$\frac{d^3y}{dx^3} = \qquad = \frac{1}{x^3}\left(\frac{d^3y}{dz^3} - 3\frac{d^2y}{dz^2} + 2\frac{dy}{dz}\right),$$

$$\cdot \quad \cdot \quad \cdot \quad \cdot \quad \cdot \quad \cdot$$

$$\frac{d^ny}{dx^n} = \qquad = \frac{1}{x^n}\left[\frac{d^ny}{dz^n} - \frac{n(n-1)}{2}\frac{d^{n-1}y}{dz^{n-1}} + \cdots + (-1)^{n-1}(n-1)!\frac{dy}{dz}\right];$$

or if we let $\frac{dy}{dz} = \mathcal{D}y$, we have

$$\begin{aligned} xDy &= \mathcal{D}y, \\ x^2D^2y &= \mathcal{D}(\mathcal{D}-1)y, \\ x^3D^3y &= \mathcal{D}(\mathcal{D}-1)(\mathcal{D}-2)y, \\ &\cdot \quad \cdot \quad \cdot \quad \cdot \quad \cdot \quad \cdot \\ x^nD^ny &= \mathcal{D}(\mathcal{D}-1)(\mathcal{D}-2)\cdots(\mathcal{D}-n+1)y; \end{aligned}$$

and (1) becomes

(2) $$[k_0\mathcal{D}(\mathcal{D}-1)\cdots(\mathcal{D}-n+1) + k_1\mathcal{D}(\mathcal{D}-1)\cdots(\mathcal{D}-n+2) + \cdots + k_{n-1}\mathcal{D} + k_n]y = Z,$$

where Z is what X becomes as a result of the transformation.*

(2) is obviously a linear equation with constant coefficients.

More generally, the equation

$$k_0(a+bx)^n\frac{d^ny}{dx^n} + k_1(a+bx)^{n-1}\frac{d^{n-1}y}{dx^{n-1}} + \cdots + k_{n-1}(a+bx)\frac{dy}{dx} + k_ny = X\dagger$$

is readily seen to be reducible to a linear equation with constant coefficients by the substitution $a + bx = e^z$.

* For another general method of solving a Cauchy linear equation see footnote, § 74.

† This form of the linear equation is referred to as *Legendre's linear equation*, after Adrien Marie Legendre (1752–1833).

Ex. 1. $x^3\dfrac{d^3y}{dx^3}+x\dfrac{dy}{dx}-y=x\log x.$

Putting $x=e^z$, this becomes

$$[\vartheta(\vartheta-1)(\vartheta-2)+\vartheta-1]y=e^z z,$$

or

$$(\vartheta^3-3\vartheta^2+3\vartheta-1)y=ze^z.$$

The roots of the auxiliary equation are 1, 1, 1.
Hence the complementary function is $Y=(c_1+c_2z+c_3z^2)e^z$.
In this case method I (§ 47) gives the particular integral at once.

We have $U=e^z\displaystyle\int\!\!\int\!\!\int e^{-z}ze^z(dz)^3=e^z\,\frac{z^4}{24}.$

$\therefore$ The solution is $y=(c_1+c_2z+c_3z^2)e^z+\dfrac{z^4e^z}{24}$,

or

$$y=[c_1+c_2\log x+c_3(\log x)^2]\,x+\frac{x\,(\log x)^4}{24}.$$

Ex. 2. $(x^3D^3+2x^2D^2+2)y=10\left(x+\dfrac{1}{x}\right).$

Ex. 3. $(x^2D^2+3xD+1)y=\dfrac{1}{(1-x)^2}.$

Ex. 4. $(x+1)^2\dfrac{d^2y}{dx^2}-4(x+1)\dfrac{dy}{dx}+6y=x.$

52. Summary.—The problem of solving a linear differential equation consists of two parts, the finding of the complementary function, and the finding of a particular integral (§ 42).

The finding of the complementary function in the case of an equation with constant coefficients $f(D)y=X$ is simply an algebraic problem, viz. the solution of the equation $f(m)=0$. According as the roots are distinct and real, repeated or complex, the complementary function takes one of the forms indicated in §§ 43, 44, 45. The problem of finding the particular integral may be attacked by any of the four methods given in §§ 47, 48, 49, 50.

An estimate of the relative merits of these methods may be summarized as follows: The methods of §§ 47 and 48 (which will be referred to as I and II respectively) and that of variation of parameters (§ 49) have the advantage of absolute generality. But as is usually true in such cases, the actual carrying out of these methods is frequently very long and laborious. Excepting in certain cases, soon learned by experience, method II is simpler than I, in that it requires several integrations of the same kind instead of several successive integrations. The method of variation of parameters has the great advantage of being readily retained in mind, but is frequently long and laborious, especially if the equation is of higher order than the second. The method of undetermined coefficients (§ 50), although not absolutely general, applies to a very large number of cases that actually occur. In such cases where it does apply it has the advantage of involving only the operations of differentiation and the solution of simultaneous linear algebraic equations. Integration is not involved. Besides, it is very readily retained in mind. The actual work of carrying out this method is straightforward and not difficult. It may at times be long, but usually it is no longer than the other methods, if as long.

As a rule, then, whenever the method of undetermined coefficients applies, it is probably the most desirable one to use. An instance of an exception to this is illustrated by Ex. 1, § 51. [Generally we may say that the method I is preferable in case the right-hand member contains a term e^{lx} or $e^{lx}f(x)$, where $f(x)$ can be integrated readily any number of times, and when the auxiliary equation is $(m - l)^r = 0$.] If it is obvious on inspection that different methods apply most readily to the various terms in the right-hand member, employ the method that is simplest for each term and take the sum of the results. This is true, for instance, in Ex. 11 below.

The equation

$$k_0(a+bx)^n\frac{d^ny}{dx^n}+k_1(a+bx)^{n-1}\frac{d^{n-1}y}{dx^{n-1}}+\cdots+k_{n-1}(a+bx)\frac{dy}{dx}+k_ny=X$$

(including as a special case the Cauchy equation where $a=0$, $b=1$) is reducible to one with constant coefficients by the substitution $a+bx=e^z$ (§ 51).

Ex. **1.** $(D^2-5D+6)y=\cos x-e^{2x}$.

Ex. **2.** $(D^4-1)y=e^x\cos x$.

Ex. **3.** $(D^2+2D+1)y=2x^3-xe^{3x}$.

Ex. **4.** $(D+1)^3y=xe^{-x}$.

Ex. **5.** $(D^3-4D)y=x^2-3e^{2x}$.

Ex. **6.** $(D^4-2D^2+1)y=\cos x$.

Ex. **7.** $(x^4D^4+6x^3D^3+9x^2D^2+3xD+1)y=(1+\log x)^2$.

Ex. **8.** $(D^3+2D^2+D)y=x^2-x$.

Ex. **9.** $(D^2+4)y=\sin^2 x$.

Ex. **10.** $(D^2+1)y=\sec^2 x$.

Ex. **11.** $(D-1)^3y=x-x^3e^x$.

Ex. **12.** $(D^4-D^3-3D^2+5D-2)y=e^{3x}$.

Ex. **13.** $(D^2+1)y=x\cos x$.

Ex. **14.** $(x^3D^3+2x^2D^2-xD+1)y=\frac{1}{x}$.

Ex. **15.** $(D^3-1)y=xe^x+\cos^2 x$.

Ex. **16.** $(D-1)^2y=\cos x+e^x+x^2e^x$.

Ex. **17.** Study the motion of a simple pendulum of length l and mass m swinging in a vacuum.

The only force acting is gravity; it acts vertically downward, and its intensity is $-mg$. If s represents the length of arc measured from the lowest point of the pendulum, then at any moment when the pendulum makes an angle θ with the vertical, $s = l\theta$, and the acceleration is $l\frac{d^2\theta}{dt^2}$. The component of the force of gravity along the tangent to the path is $-mg \sin \theta$. Hence the equation of motion is

$$ml\frac{d^2\theta}{dt^2} = -mg \sin \theta.$$

If θ remains very small throughout the motion, we may replace $\sin \theta$ by θ as a first approximation. Our equation then takes the form

$$\frac{d^2\theta}{dt^2} + \frac{g}{l}\theta = 0.$$

[This is the differential equation of simple harmonic motion.]

Solving this, we have $\theta = A \cos\left(\sqrt{\frac{g}{l}}t + B\right)$,

where A and B are constants depending on the initial value of θ and of $\frac{d\theta}{dt}$.

A determines the amplitude, while B determines the phase.

The period is $2\pi\sqrt{\frac{l}{g}}$, *i.e.* the state of motion will be identically the same for two values of t whose difference is an integral multiple of this quantity.

Ex. 18. Consider the case of a simple pendulum moving in a resisting medium where the resisting force is proportional to the velocity, say $-2\,km\frac{ds}{dt}$.

Putting $\frac{g}{l} = n^2$, the differential equation to be solved is

$$\frac{d^2\theta}{dt^2} + 2k\frac{d\theta}{dt} + n^2\theta = 0.$$

[The same equation arises in the case of damped vibrations of the needle of a galvanometer.]

Ex. **19.** In the case of forced vibrations, such as when a magnet is brought up periodically to a vibrating tuning fork, the equation of motion, in case there is no resisting force, is

$$(a) \qquad \frac{d^2\theta}{dt^2} + n^2\theta = C \cos mt,$$

(the cases when $m \neq n$ and $m = n$ must be distinguished).

If the resisting force is proportional to the velocity, the equation of motion is

$$(b) \qquad \frac{d^2\theta}{dt^2} + 2k\frac{d\theta}{dt} + n^2\theta = C \cos mt.$$

Ex. **20.** A particle is projected with velocity v_0 away from the center of an attractive force. If the acceleration of the particle due to the force is proportional to the distance, find the motion.

Ex. **21.** If in Ex. 20, the force is a repellent one, and the particle is projected toward the center of force with the velocity v_0, find the motion.

Ex. **22.** Find the motion of a heavy particle moving without friction along a massless straight line which rotates about one of its points in a vertical plane with constant angular velocity. The only force acting is gravity.

If r is the distance of the moving point from the point about which the line rotates, and if ω is the angular velocity of the line, Lagrange's equation (generalized coördinates) is

$$\frac{d^2r}{dt^2} - \omega^2 r = -g \sin \omega t.$$

Ex. **23.** If a condenser of capacity S, charged with a quantity of electricity Q, is introduced into an electric circuit, it will discharge by sending a current through the circuit. If q is the quantity of electricity in the condenser at any instant during the discharge, it will be determined by

$$\frac{d^2q}{dt^2} + \frac{R}{L}\frac{dq}{dt} + \frac{q}{LS} = 0,$$

where L is the self-inductance and R the resistance of the circuit.

Here the auxiliary equation is

$$m^2 + \frac{R}{L}m + \frac{1}{LS} = 0.$$

$$\therefore m_1, m_2 = -\frac{R}{2L} \pm \sqrt{\frac{R^2}{4L^2} - \frac{1}{LS}}.$$

1° If $$R^2 > \frac{4L}{S},$$

$$q = Ae^{m_1 t} + Be^{m_2 t}.$$

To determine A and B we make use of the fact that when $t = 0$, $q = Q$, and $-\frac{dq}{dt} \equiv i = 0$, where i is the current; *i.e.*

$$A + B = Q, \text{ and } m_1 A + m_2 B = 0,$$

whence $$A = -\frac{m_2 Q}{m_1 - m_2},\ B = \frac{m_1 Q}{m_1 - m_2}, \text{ and}$$

$$q = \frac{Q}{m_1 - m_2}(m_1 e^{m_2 t} - m_2 e^{m_1 t}),$$

$$i = -\frac{dq}{dt} = \frac{m_1 m_2 Q}{m_1 - m_2}(e^{m_1 t} - e^{m_2 t}).$$

Noting the values of m_1 and m_2, we see that q and i diminish continually, but do not become zero for a finite value of t, although they are practically negligible very soon when $\frac{R}{L}$ is a large quantity, which it usually is.

2° If $$R^2 = \frac{4L}{S},$$

$$q = e^{-\frac{R}{2L}t}(A + Bt).$$

To determine A and B we have,

$$A = Q,\ \frac{R}{2L}A - B = 0,\ \text{ or } B = \frac{QR}{2L},$$

whence $$q = \frac{Q}{2L}(2L + Rt)e^{-\frac{R}{2L}t},$$

$$i = -\frac{dq}{dt} = \frac{QR^2 t}{4L^2}e^{-\frac{R}{2L}t}.$$

Here again q and i diminish rapidly, without vanishing for a finite value of t, although they are soon negligible as a rule.

3° If $R^2 < \frac{4L}{S}$, $m_1, m_2 = -\frac{R}{2L} \pm i\sqrt{\frac{1}{LS} - \frac{R^2}{4L^2}}$

$$= \alpha \pm i\beta.$$

$\therefore\ q = e^{\alpha t}(A \cos \beta t + B \sin \beta t)$, and

$$i = -\frac{dq}{dt} = -e^{\alpha t}[(\alpha A + \beta B)\cos \beta t + (\alpha B - \beta A)\sin \beta t].$$

To determine A and B we have,

$$A = Q,\ \alpha A + \beta B = 0, \text{ or } B = -\frac{\alpha Q}{\beta},$$

whence

$$q = Qe^{-\frac{R}{2L}t}\left[\cos t\sqrt{\frac{1}{LS} - \frac{R^2}{4L^2}} + \frac{\frac{R}{2L}}{\sqrt{\frac{1}{LS} - \frac{R^2}{4L^2}}}\sin t\sqrt{\frac{1}{LS} - \frac{R^2}{4L^2}}\right],$$

$$i = \frac{Q}{LS\sqrt{\frac{1}{LS} - \frac{R^2}{4L^2}}}e^{-\frac{R}{2L}t}\sin t\sqrt{\frac{1}{LS} - \frac{R^2}{4L^2}}.$$

Both q and i are periodic functions of period $T = \dfrac{2\pi}{\sqrt{\frac{1}{LS} - \frac{R^2}{4L^2}}}$, so that sometimes they are positive and sometimes negative. The amplitude in either case is a constant times $e^{-\frac{R}{2L}t}$, which usually diminishes very rapidly with t. But in specially constructed circuits in which R is small relative to L, an oscillatory discharge may be realized.—I. C. and J. P. JACKSON, *Alternating Currents and Alternating Current Machinery.*

CHAPTER VIII

LINEAR DIFFERENTIAL EQUATIONS OF THE SECOND ORDER*

53. Change of Dependent Variable. — While the problem of solving linear differential equations of the first order can always be carried out (that is to say, we can reduce it to one of quadratures, § 13), that of solving equations of the second order can be carried out in only a comparatively small number of cases.

The general type of a linear equation of the second order is

$$\frac{d^2y}{dx^2} + P\frac{dy}{dx} + Qy = X, \tag{1}$$

where P, Q, X are functions of x only.

Let us try the following change of dependent variable,

$$y = y_1 v. \tag{2}$$

Then $\dfrac{dy}{dx} = y_1\dfrac{dv}{dx} + \dfrac{dy_1}{dx}v, \quad \dfrac{d^2y}{dx^2} = y_1\dfrac{d^2v}{dx^2} + 2\dfrac{dy_1}{dx}\dfrac{dv}{dx} + \dfrac{d^2y_1}{dx^2}v;$

and equation (1) becomes

$$\frac{d^2v}{dx^2} + P_1\frac{dv}{dx} + Q_1 v = X_1, \tag{3}$$

* In this chapter we shall consider methods which apply more especially to linear differential equations of the second order. Of course, the general methods of the next chapter apply to linear equations of the second order as well. But owing to the general plan of solution of equations of higher order than the first (§ 56), it is desirable to have available the methods given in this chapter.

where $$P_1 = \frac{2}{y_1}\frac{dy_1}{dx} + P, \quad Q_1 = \frac{\dfrac{d^2y_1}{dx^2} + P\dfrac{dy_1}{dx} + Qy_1}{y_1}, \quad X_1 = \frac{X}{y_1}.$$

Two uses may be made of this.

1° By inspection* or other means a particular integral of the equation when we put $X = 0$ may be known. If we let this be y_1, we have $Q_1 = 0$, so that (3) becomes

$$\frac{d^2v}{dx^2} + P_1\frac{dv}{dx} = X_1.$$

If now we let $\dfrac{dv}{dx} = p$, we have a linear equation of the first order

$$\frac{dp}{dx} + P_1 p = X_1,$$

which can be solved for p (§ 13). A quadrature will then give y.

Ex. 1. $\dfrac{d^2y}{dx^2} - x^2\dfrac{dy}{dx} + xy = x.$

Here x is a particular integral (since $P = -Qx$). Putting $y = xv$, we have

$$x\frac{d^2v}{dx^2} + (2 - x^3)\frac{dv}{dx} = x, \text{ or, putting } \frac{dv}{dx} = p,$$

$$\frac{dp}{dx} + \left(\frac{2}{x} - x^2\right)p = 1.$$

* Thus, for example, if $P = -Qx$, x is evidently such a particular integral. Again, if $1 + P + Q = 0$, e^x is such, or, if $1 - P + Q = 0$, e^{-x} is one; or more generally, it may be possible to note, by inspection, a number m, such that $m^2 + Pm + Q = 0$; in this case e^{mx} is such an integral.

An integrating factor is $e^{\int\left(\frac{2}{x}-x^2\right)dx}$ or $x^2e^{-\frac{x^3}{3}}$.

$$\therefore\ px^2e^{-\frac{x^3}{3}}=\int x^2e^{-\frac{x^3}{3}}dx=-e^{-\frac{x^3}{3}}+c_1,$$

or
$$p=\frac{dv}{dx}=-\frac{1}{x^2}+c_1x^{-2}e^{\frac{x^3}{3}},$$

and
$$v=\frac{1}{x}+c_1\int x^{-2}e^{\frac{x^3}{3}}dx+c_2;$$

whence
$$y=1+c_1x\int x^{-2}e^{\frac{x^3}{3}}dx+c_2x.$$

Ex. 2. $x\dfrac{d^2y}{dx^2}-(2x+1)\dfrac{dy}{dx}+(x+1)y=x^2-x-1.$

Ex. 3. $(1+x^2)\dfrac{d^2y}{dx^2}+2x\dfrac{dy}{dx}-2y=0.$

Ex. 4. $(1-x)\dfrac{d^2y}{dx^2}+x\dfrac{dy}{dx}-y=(1-x)^2.$

Here x and e^x are particular integrals, when the right-hand member is replaced by zero. Hence, by property A, § 42, the complementary function is $c_1x+c_2e^x$. To find the particular integral which must be added to the complementary function, the method of variation of parameters (§ 49) may be employed.*

* Besides the method of variation of parameters one can sometimes use with facility a general form for the particular integral given by Lie in his *Differentialgleichungen*, p. 429. This form also appears in the author's *Lie Theory of One-Parameter Groups*, p. 174.

2° If we put $P_1 = 0$, *i.e.* $\frac{2}{y_1}\frac{dy_1}{dx} + P = 0$,

we have
$$\log y_1 = -\frac{1}{2}\int P\,dx, \text{ or}$$

(4)
$$y_1 = e^{-\frac{1}{2}\int P dx}.$$

Using this value for y_1, we have

$$Q_1 = Q - \frac{1}{2}\frac{dP}{dx} - \frac{1}{4}P^2, \quad X_1 = Xe^{\frac{1}{2}\int P dx}.$$

Now it may turn out that $Q - \frac{1}{2}\frac{dP}{dx} - \frac{1}{4}P^2$ is a constant, in which case (3) is an equation with constant coefficients; or it may be a constant divided by x^2, in which case we have a Cauchy equation, and the further substitution $x = e^t$ will reduce it to one with constant coefficients (§ 51).

Ex. 5. $\sin x \frac{d^2y}{dx^2} + 2\cos x \frac{dy}{dx} + 3\sin x \cdot y = e^x$,

or
$$\frac{d^2y}{dx^2} + 2\cot x \frac{dy}{dx} + 3y = e^x \csc x.$$

Here
$$Q - \frac{1}{2}\frac{dP}{dx} - \frac{1}{4}P^2 = 3 + \csc^2 x - \cot^2 x = 4.$$

Hence $y = ve^{-\frac{1}{2}\int P dx} = v \csc x$ transforms the equation to

$$\frac{d^2v}{dx^2} + 4v = e^x.\text{*}$$

Integrating,
$$v = c_1 \cos 2x + c_2 \sin 2x + \frac{1}{5}e^x,$$

and
$$y = c_1(\cos x \cot x - \sin x) + c_2 \cos x + \frac{1}{5}e^x \csc x.$$

* This result can be written at once without actually carrying out the transformation.

Ex. 6. $\frac{d^2y}{dx^2} - 2\tan x\frac{dy}{dx} - (a^2+1)y = 0.$

Ex. 7. $4x^2\frac{d^2y}{dx^2} + 4x^3\frac{dy}{dx} + (x^2+1)^2y = 0.$

Ex. 8. $x\frac{d^2y}{dx^2} + 2\frac{dy}{dx} - xy = 2e^x.$

54. Change of Independent Variable.—If we introduce a new independent variable z, we have

$$\frac{dy}{dx} = \frac{dy}{dz}\frac{dz}{dx}, \quad \frac{d^2y}{dx^2} = \frac{d^2y}{dz^2}\left(\frac{dz}{dx}\right)^2 + \frac{dy}{dz}\frac{d^2z}{dx^2},$$

and the equation (1) becomes

$$(5) \qquad \frac{d^2y}{dz^2} + \frac{\frac{d^2z}{dx^2} + P\frac{dz}{dx}}{\left(\frac{dz}{dx}\right)^2}\frac{dy}{dz} + \frac{Q}{\left(\frac{dz}{dx}\right)^2}y = \frac{X}{\left(\frac{dz}{dx}\right)^2}.$$

It may happen that if we put $\frac{Q}{\left(\frac{dz}{dx}\right)^2} = \pm 1$, *i.e.* $\frac{dz}{dx} = \sqrt{\pm Q}$ (where we choose that sign which will make the square root real), the coefficient of $\frac{dy}{dz}$ reduces to a constant. If such is the case, our equation (5) is linear with constant coefficients, and can be solved by the methods of Chapter VII.

Remark. — If the result of putting $\dfrac{Q}{\left(\dfrac{dz}{dx}\right)^2} = \pm 1$ is to transform the equation into $\dfrac{d^2y}{dz^2} + K\dfrac{dy}{dz} \pm y = 0$, the transformation $\dfrac{Q}{\left(\dfrac{dz}{dx}\right)^2} = \pm a$, where a is any constant, will give us $\dfrac{d^2y}{dz^2} + \sqrt{a}\,K\dfrac{dy}{dz} \pm ay = 0$. In either case we have a linear equation with constant coefficients. But if K involves a square root factor, a may be so chosen that $\sqrt{a}\,K$ is rational, and the actual work is thus simplified. For example, see Ex. 5 below.

Ex. 1. $\dfrac{d^2y}{dx^2} + (2e^x - 1)\dfrac{dy}{dx} + e^{2x}y = e^{4x}.$

If
$$\frac{dz}{dx} = e^x, \quad \frac{\dfrac{d^2z}{dx^2} + (2e^x - 1)\dfrac{dz}{dx}}{\left(\dfrac{dz}{dx}\right)^2} = 2.$$

Hence, introducing $z = e^x$ as the new independent variable, the equation becomes
$$\frac{d^2y}{dz^2} + 2\frac{dy}{dz} + y = z^2.\text{*}$$

Its solution is
$$y = (c_1 + c_2z)e^{-z} + z^2 - 4z + 6.$$

Replacing z by its value in terms of x, we have finally
$$y = (c_1 + c_2e^x)e^{-e^x} + e^{2x} - 4e^x + 6.$$

Ex. 2. $(1 - x^2)\dfrac{d^2y}{dx^2} - x\dfrac{dy}{dx} + 4y = 0.$

Ex. 3. $\dfrac{d^2y}{dx^2} + \tan x\dfrac{dy}{dx} + \cos^2 x \cdot y = 0.$

Ex. 4. $x^6\dfrac{d^2y}{dx^2} + 3x^5\dfrac{dy}{dx} + y = \dfrac{1}{x^2}.$

Ex. 5. $x\dfrac{d^2y}{dx^2} - (2x^2 + 1)\dfrac{dy}{dx} - 8x^3y = 4x^3e^{-x^2}.$

* This result can be written at once without carrying out the transformation.

55. Summary. — There is no general method for solving the linear differential equation of the second order, $\frac{d^2y}{dx^2} + P\frac{dy}{dx} + Qy = X$. In actual practice we proceed as follows:

1°. If by inspection, or otherwise, we know a particular integral y_1 when the right-hand member is made zero, then $y = y_1 v$ will reduce the equation to a linear one of the first order when $\frac{dv}{dx}$ is considered a new variable (§ 53, 1°).

2° If such a particular integral is not known, the next thing to do is to find the value of $Q - \frac{1}{2}\frac{dP}{dx} - \frac{1}{4}P^2$. If this is a constant (or a constant divided by x^2) the equation is reducible to one with constant coefficients (or to a Cauchy equation) by substituting $y = y_1 v$, ~~and then it is time to calculate~~ $y_1 = e^{-\frac{1}{2}\int P\,dx}$ (§ 53, 2°).*

3° If the previous method does not apply, put $\frac{dz}{dx} = \sqrt{\pm Q}$ (using that sign which will make the square root real); then substitute in

$$\frac{\frac{d^2z}{dx^2} + P\frac{dz}{dx}}{\left(\frac{dz}{dx}\right)^2}.$$

If this turns out to be a constant, the method applies, and then it is time to find z from $\frac{dz}{dx} = \sqrt{\pm Q}$ (§ 54).*

Ex. 1. $x\frac{d^2y}{dx^2} - (x+3)\frac{dy}{dx} + 3y = 0.$

Ex. 2. $(x-3)\frac{d^2y}{dx^2} - (4x-9)\frac{dy}{dx} + (3x-6)y = 0.$

Ex. 3. $x^2\frac{d^2y}{dx^2} + 4x\frac{dy}{dx} + (2 - x^2)y = 0.$

* Emphasis should be laid on the fact that in the application of the test as to whether this method applies no integration is required. It is only after one is assured the method works that a new variable need be sought.

Ex. 4. $(x^2+1)\frac{d^2y}{dx^2} - 2x\frac{dy}{dx} + 2y = 0.$

Ex. 5. $x\frac{d^2y}{dx^2} - (2x-1)\frac{dy}{dx} + (x-1)y = 0.$

Ex. 6. $x^2\frac{d^2y}{dx^2} - 4x\frac{dy}{dx} + (6+x^2)y = 0.$

Ex. 7. $(2x^3-1)\frac{d^2y}{dx^2} - 6x^2\frac{dy}{dx} + 6xy = 0.$

Ex. 8. $x^2\frac{d^2y}{dx^2} - 2x(1+x)\frac{dy}{dx} + 2(1+x)y = x^3.$

Ex. 9. $x^2\frac{d^2y}{dx^2} - 2nx\frac{dy}{dx} + (n^2+n+a^2x^2)y = 0.$

Ex. 10. $x^4\frac{d^2y}{dx^2} + 2x^3\frac{dy}{dx} + n^2y = 0.$

CHAPTER IX

MISCELLANEOUS METHODS FOR SOLVING EQUATIONS OF HIGHER ORDER THAN THE FIRST

56. General Plan of Solution. — There is no general direct method for solving a differential equation of higher order than the first, excepting in the case of linear equations with constant coefficients and those reducible to such (Chapter VII). The general plan in all other cases is to try to transfer the problem to that of solving an equation of lower order. We shall consider some classes of equations for which this can be done.

57. Dependent Variable Absent. — If y is absent, the equation is of the form

$$f\left(\frac{d^ny}{dx^n}, \frac{d^{n-1}y}{dx^{n-1}}, \cdots, \frac{d^2y}{dx^2}, \frac{dy}{dx}, x\right)=0.$$

If we put $\frac{dy}{dx}=p$, then

$$\frac{d^2y}{dx^2}=\frac{dp}{dx}, \cdots, \frac{d^ny}{dx^n}=\frac{d^{n-1}p}{dx^{n-1}},$$

and the equation to be solved is

$$f\left(\frac{d^{n-1}p}{dx^{n-1}}, \frac{d^{n-2}p}{dx^{n-2}}, \cdots, \frac{dp}{dx}, p, x\right)=0,$$

which is of order $n-1$. If this can be solved for p, we have $p=\phi(x, c_1, c_2, \cdots, c_{n-1})$, and y can be obtained by the quadrature

$$y=\int\phi(x, c_1, c_2, \cdots, c_{n-1})dx+c_n.$$

More generally, if y and all of its derivatives up to the $(r-1)$st are absent, so that the equation is of the form

$$f\left(\frac{d^ny}{dx^n}, \cdots, \frac{d^{r+1}y}{dx^{r+1}}, \frac{d^ry}{dx^r}, x\right)=0,$$

by letting $\frac{d^ry}{dx^r}=v$, the equation becomes

$$f\left(\frac{d^{n-r}v}{dx^{n-r}}, \cdots, \frac{dv}{dx}, v, x\right)=0,$$

which is of order $n-r$. If this can be solved for v, we have $v=\phi(x, c_1, c_2, \cdots, c_{n-r})$, and y can be obtained by r successive quadratures, *i.e.* $y=\int\int\cdots\int\phi(x, c_1, c_2, \cdots, c_{n-r})dx^r + c_{n-r+1}x^{r-1}$
$+ c_{n-r+2}x^{r-2} + \cdots + c_{n-1}x + c_n.$

If y and all of its derivatives except the highest are absent, the equation may be put in the form

$$\frac{d^ny}{dx^n}=f(x),$$

and the solution is obtained directly by n successive quadratures. For $\frac{d^{n-1}y}{dx^{n-1}}=\int f(x)dx + a_1$, whence $\frac{d^{n-2}y}{dx^{n-2}}=\int\int f(x)dx^2 + a_1x + a_2$, and so on, until we get

$$y=\int\int\cdots\int f(x)dx^n + c_1x^{n-1} + c_2x^{n-2} + \cdots + c_{n-1}x + c_n.$$

Ex. 1. $(1+x^2)\frac{d^2y}{dx^2}+1+\left(\frac{dy}{dx}\right)^2=0.$

Putting $\dfrac{dy}{dx} = p$, we have

$$(1 + x^2)\frac{dp}{dx} + 1 + p^2 = 0, \text{ or } \frac{dp}{1 + p^2} + \frac{dx}{1 + x^2} = 0.$$

$$\therefore \tan^{-1} p = c - \tan^{-1} x, \text{ or } p = \frac{c_1 - x}{1 + c_1 x}, \text{ where } c_1 = \tan c.$$

Integrating, we have $c_1^2 y = (c_1^2 + 1)\log(1 + c_1 x) - c_1 x + c_2$.

Ex. 2. $\left(x\dfrac{d^3y}{dx^3} - \dfrac{d^2y}{dx^2}\right)^2 = \left(\dfrac{d^3y}{dx^3}\right)^2 + 1.$

Putting $\dfrac{d^2y}{dx^2} = v$ and solving for this, we have

$$v = x\frac{dv}{dx} \pm \sqrt{\left(\frac{dv}{dx}\right)^2 + 1}.$$

This is Clairaut's form (§ 27) and has for solution

$$v = \frac{d^2y}{dx^2} = cx \pm \sqrt{c^2 + 1}.$$

Integrating, we get $\qquad \dfrac{dy}{dx} = \dfrac{c}{2}x^2 \pm x\sqrt{c^2 + 1} + c'.$

Integrating again, $\qquad y = \dfrac{c}{6}x^3 \pm \dfrac{x^2}{2}\sqrt{c^2 + 1} + c'x + c''.$

Ex. 3. $\dfrac{d^2y}{dx^2} + x\dfrac{dy}{dx} = x.$

Ex. 4. $\dfrac{d^2y}{dx^2} = xe^x.$

Ex. 5. $\left(\dfrac{dy}{dx} - x\dfrac{d^2y}{dx^2}\right)^2 = 1 + \left(\dfrac{d^2y}{dx^2}\right)^2.$

58. Independent Variable Absent. — If x is absent, by taking y as the independent variable and letting $\frac{dy}{dx}, \equiv p$, be the dependent one, we have

$$\frac{d^2y}{dx^2} = p\frac{dp}{dy},$$

$$\frac{d^3y}{dx^3} = p^2\frac{d^2p}{dy^2} + p\left(\frac{dp}{dy}\right)^2,$$

$$\frac{d^4y}{dx^4} = p^3\frac{d^3p}{dy^3} + 4p^2\frac{dp}{dy}\frac{d^2p}{dy^2} + p\left(\frac{dp}{dy}\right)^3,$$

$$\frac{d^5y}{dx^5} = p^4\frac{d^4p}{dy^4} + 7p^3\frac{dp}{dy}\frac{d^3p}{dy^3} + 11p^2\left(\frac{dp}{dy}\right)^2\frac{d^2p}{dy^2} + 4p^3\left(\frac{d^2p}{dy^2}\right)^2 + p\left(\frac{dp}{dy}\right)^4,$$

. . . .

and the equation becomes one of order $n - 1$. If this can be solved for p, we have $p = \phi(y, c_1, c_2, \cdots, c_{n-1})$, and y can be obtained by the quadrature $\int \frac{dy}{\phi(y, c_1, c_2, \cdots, c_{n-1})} = x + c_n$.

Remark. — The equation $\frac{d^2y}{dx^2} = f(y)$, which belongs to the class of equations here considered, has the obvious integrating factor $2\frac{dy}{dx}dx$. Using it, we have

$$2\frac{dy}{dx}\frac{d^2y}{dx^2}dx = 2f(y)\,dy.$$

Integrating, we get $$\left(\frac{dy}{dx}\right)^2 = 2\int f(y)\,dy + c_1;$$

whence $$\int \frac{dy}{\sqrt{2\int f(y)\,dy + c_1}} = x + c_2.$$

It should be noted that the general method of this paragraph leads to exactly this method of solution.

Ex. 1. $y\frac{d^2y}{dx^2}-\left(\frac{dy}{dx}\right)^2-y^2\frac{dy}{dx}=0.$

Putting $\frac{dy}{dx}=p, \frac{d^2y}{dx^2}=p\frac{dp}{dy}$, we get

$$p\left(y\frac{dp}{dy}-p-y^2\right)=0.$$

The factor $p'=0$ gives $y=c$, a particular solution.

$y\frac{dp}{dy}-p-y^2=0$ has the obvious integrating factor $\frac{1}{y^2}$. Using this we have $\frac{p}{y}=y+c$. Remembering that $p=\frac{dy}{dx}$, we have

$$\frac{dy}{y(y+c)}=dx,$$

whence
$$\log\frac{y}{y+c}=cx+c';$$

$$\text{or } \frac{y}{y+c}=ke^{cx}.$$

Ex. 2. $y\frac{d^2y}{dx^2}+\left(\frac{dy}{dx}\right)^2+1=0.$

Ex. 3. $2\frac{d^2y}{dx^2}=e^y.$

Ex. 4. $y\frac{d^2y}{dx^2}+2\frac{dy}{dx}-\left(\frac{dy}{dx}\right)^2=0.$

59. Linear Equations with Particular Integral Known. — If the equation is linear and of any order, and a particular integral is known when the right-hand member is made zero, the method 1°, § 53, applies.* Thus, let the equation be

$$(P_0D^n+P_1D^{n-1}+\cdots+P_{n-1}D+P_n)y=X.$$

* The hint there given as to how a particular integral may at times be found, applies equally well here.

Putting $y = y_1 v$, we have $Dy = Dy_1 \cdot v + \cdots$,*

$$D^2 y = D^2 y_1 \cdot v + \cdots,$$

$$\cdot \quad \cdot \quad \cdot \quad \cdot$$

$$D^n y = D^n y_1 \cdot v + \cdots.$$

Making the substitution, we have

$$\cdots + (P_0 D^n + P_1 D^{n-1} + \cdots + P_n) y_1 \cdot v = X.$$

By hypothesis, the coefficient of v is zero. Hence, on letting $\frac{dv}{dx} = p$, the equation reduces to one of order $n - 1$.

Ex. 1. $[(x^2 - 2x + 2)D^3 - x^2 D^2 + 2xD - 2]y = 0.$

$y = x$ is a particular solution. Putting $y = xv$, we have

$$(x^3 - 2x^2 + 2x) D^3 v - (x^3 - 3x^2 + 6x - 6) D^2 v = 0.$$

Letting $D^2 v = q$, this becomes

$$\frac{dq}{q} = \frac{x^3 - 3x^2 + 6x - 6}{x^3 - 2x^2 + 2x} dx = dx - \frac{3\,dx}{x} + \frac{2x - 2}{x^2 - 2x + 2} dx.$$

$$\log q = x - 3 \log x + \log (x^2 - 2x + 2) + c.$$

$$\therefore q = \frac{d^2 v}{dx^2} = c_1 e^x \frac{x^2 - 2x + 2}{x^3}.$$

Integrating, $$\frac{dv}{dx} = c_1 e^x \frac{x - 1}{x^2} + c_2,$$

and $$v = c_1 e^x \frac{1}{x} + c_2 x + c_3.$$

Therefore $$y = c_1 e^x + c_2 x^2 + c_3 x.$$

* ... stands for terms free of v, and involving its derivatives to an order as high as the exponent of D on the left.

Ex. 2. $(xD^3 - D^2 - xD + 1)y = 1 - x^2$.

By inspection, it is seen that e^x, e^{-x}, x are particular integrals, hence we know at once that the complementary function is

$$Y = c_1 e^x + c_2 e^{-x} + c_3 x.$$

The student should verify that, by the method of variation of parameters (§ 49), this becomes the general solution when

$$c_1 = \frac{e^{-x}}{2}(x + 2 - x^{-1}) + k_1,$$

$$c_2 = \frac{e^x}{2}(x^{-1} + 2 - x) + k_2,$$

$$c_3 = x + \frac{1}{x} + k_3.$$

Hence the solution is

$$y = k_1 e^x + k_2 e^{-x} + k_3 x + x^2 + 3.$$

In order to get practice in the general method of this paragraph, let the student solve this example by that method.

60. Exact Equation. Integrating Factor. — In case the equation is the derivative of another one, the order may be reduced by direct integration. No simple formula can be given as a test for exactness (except in the case of linear equations). But the method is simple and direct, and can probably be brought out best by the following examples:

Consider first the linear equation,

$$(1) \qquad P_0 \frac{d^3y}{dx^3} + P_1 \frac{d^2y}{dx^2} + P_2 \frac{dy}{dx} + P_3 y = X.$$

$P_0 \frac{d^3y}{dx^3}$ will arise on differentiating $P_0 \frac{d^2y}{dx^2}$. But differentiating this, we get $P_0 \frac{d^3y}{dx^3} + P'_0 \frac{d^2y}{dx^2}$, (indicating differentiation by a prime). Now, if (1) is exact, so is

$$(2) \qquad (P_1 - P'_0) \frac{d^2y}{dx^2} + P_2 \frac{dy}{dx} + P_3 y.$$

$(P_1 - P_0') \frac{d^2y}{dx^2}$ will arise on differentiating $(P_1 - P'_0) \frac{dy}{dx}$. But differentiating this, we get $(P_1 - P'_0) \frac{d^2y}{dx^2} + (P'_1 - P''_0) \frac{dy}{dx}$.

Hence if (2) is exact, so also is

$$(3) \qquad (P_2 - P'_1 + P''_0) \frac{dy}{dx} + P_3 y.$$

$(P_2 - P'_1 + P''_0) \frac{dy}{dx}$ will arise on differentiating $(P_2 - P'_1 + P''_0) y$.

But differentiating this, we get

$$(P_2 - P'_1 + P''_0) \frac{dy}{dx} + (P'_2 - P''_1 + P'''_0) y,$$

hence, if (3) is exact, we must have $P_3 - P'_2 + P''_1 - P'''_0 \equiv 0$.*

Moreover, this condition is also obviously sufficient, and we have that a first integral of (1) is

$$P_0 \frac{d^2y}{dx^2} + (P_1 - P'_0) \frac{dy}{dx} + (P_2 - P'_1 + P''_0) y = \int X\, dx + c.$$

* This suggests the condition for exactness of a linear equation of the nth order,

$$(4) \qquad P_n - P'_{n-1} + P''_{n-2} - \cdots + (-1)^r P^{(r)}_{n-r} + \cdots + (-1)^n P^{(n)}_0 \equiv 0.$$

This method applies also to equations that are not linear, but in such cases there is no simple test for exactness; one must actually carry out the work of finding the first integral to find out whether it is exact. Thus consider the equation

$$(y^2+x)\frac{d^3y}{dx^3}+6y\frac{dy}{dx}\frac{d^2y}{dx^2}+\frac{d^2y}{dx^2}+2\left(\frac{dy}{dx}\right)^3=0.$$

The derivative of $(y^2+x)\frac{d^2y}{dx^2}$ is $(y^2+x)\frac{d^3y}{dx^3}+2y\frac{dy}{dx}\frac{d^2y}{dx^2}+\frac{d^2y}{dx^2}$.

Subtracting this, we have $4y\frac{dy}{dx}\frac{d^2y}{dx^2}+2\left(\frac{dy}{dx}\right)^3$.

This is the derivative of $2y\left(\frac{dy}{dx}\right)^2$. Hence, a first integral is

$$(y^2+x)\frac{d^2y}{dx^2}+2y\left(\frac{dy}{dx}\right)^2=c_1.$$

Let the student show that this is also exact.

Remark. — Since an exact differential results from differentiating an expression of one lower order, it is obviously necessary that in it the highest ordered derivative appear to the first degree only. In other words, we can never expect an expression, in which the highest derivative entering appears to a higher degree than the first, to be exact. Moreover, this must be true of all the expressions [such as (2) and (3) above] which arise in the course of the process. If any one of these turns out to be of higher degree than the first, there is no need to proceed farther.

Ex. 1. $(x+2)^2\frac{d^3y}{dx^3}+(x+2)\frac{d^2y}{dx^2}+\frac{dy}{dx}=1.$

This is linear and satisfies the condition (4) for exactness.

The derivative of $(x+2)^2\frac{d^2y}{dx^2}$ is $(x+2)^2\frac{d^3y}{dx^3}+2(x+2)\frac{d^2y}{dx^2}$.

Subtracting, we get $-(x+2)\frac{d^2y}{dx^2}+\frac{dy}{dx}$.

The derivative of $-(x+2)\frac{dy}{dx}$ is $-(x+2)\frac{d^2y}{dx^2}-\frac{dy}{dx}$. Subtracting, we get $2\frac{dy}{dx}$, whose integral is $2y$. Hence, a first integral is

$$(x+2)^2\frac{d^2y}{dx^2}-(x+2)\frac{dy}{dx}+2y=x+c.$$

Since $2+1+2\neq 0$, this equation is not exact.

But putting $x+2=e^z$ (§ 51), we have

$$\frac{d^2y}{dz^2}-2\frac{dy}{dz}+2y=e^z+c',$$

a linear equation with constant coefficients.

The roots of the auxiliary equation are $1\pm i$.

Hence $Y=e^z(A\cos z+B\sin z)$ is the complementary function. For the particular integral try $U=ae^z+b$. Substituting in the equation, we must have $ae^z+2b\equiv e^z+c'$. Hence $a=1$, $b=\frac{c'}{2}$. And the solution is $y=e^z(A\cos z+B\sin z)+e^z+\frac{c'}{2}$, or

$$y=(x+2)\left[A\cos\log(x+2)+B\sin\log(x+2)\right]+x+C.$$

Remark. — It may be noted that since y is absent in the original differential equation, the method of § 57 applies. The student should solve the problem from this point of view.

Ex. 2. $x^2\frac{d^2y}{dx^2}+3x\frac{dy}{dx}+y=x.$

Ex. 3. $(x-1)^2\frac{d^2y}{dx^2}+4(x-1)\frac{dy}{dx}+2y=\cos x.$

Ex. 4. $(x^3 - x)\dfrac{d^3y}{dx^3} + (8x^2 - 3)\dfrac{d^2y}{dx^2} + 14x\dfrac{dy}{dx} + 4y = 0.$

Ex. 5. $$2x^3y\frac{d^3y}{dx^3} + 6x^3\frac{dy}{dx}\frac{d^2y}{dx^2} + 18x^2y\frac{d^2y}{dx^2} + 18x^2\left(\frac{dy}{dx}\right)^2 + 36xy\frac{dy}{dx} + 6y^2 = 0.$$

It is at times possible to find an *integrating factor.* But no general treatment of this part of the subject can be given here.* In the cases to be considered here, special methods, or inspection, will be employed. One important type of equation, arising in physical problems, has already been mentioned (Remark, § 58).

Ex. 6. $x^5\dfrac{d^2y}{dx^2} + (2x^4 - x)\dfrac{dy}{dx} - (2x^3 - 1)y = 0.$

This equation is not exact, since $-2x^3 + 1 - 8x^3 + 1 + 20x^3 \not\equiv 0.$ But x^m will be an integrating factor provided we can find a value for m such that

$$-2x^{m+3} + x^m - 2(m+4)x^{m+3} + (m+1)x^m + (m+5)(m+4)x^{m+3} \equiv 0,$$

or $$(m^2 + 7m + 10)x^{m+3} + (m+2)x^m \equiv 0;$$

i.e. $$m^2 + 7m + 10 = 0, \text{ and } m + 2 = 0.$$

Both of these will be satisfied if $m = -2$.

* For integrating factors in the case of linear equations, see Schlesinger, *Differentialgleichungen*, p. 147, and references given there.

Hence x^{-2} is an integrating factor. Using it, we have

$$x^3\frac{d^2y}{dx^2}+(2x^2-x^{-1})\frac{dy}{dx}-(2x-x^{-2})y=0.$$

$$\begin{array}{l|l}
x^3\dfrac{dy}{dx} & x^3\dfrac{d^2y}{dx^2}+3x^2\dfrac{dy}{dx} \\ \hline
 & -(x^2+x^{-1})\dfrac{dy}{dx}-(2x-x^{-2})y \\
-(x^2+x^{-1})y & -(x^2+x^{-1})\dfrac{dy}{dx}-(2x-x^{-2})y \\ \hline
\end{array}$$

Hence a first integral is $x^3\dfrac{dy}{dx}-(x^2+x^{-1})y=c$, or

$$\frac{dy}{dx}-(x^{-1}+x^{-4})y=cx^{-3}.$$

This is linear. An integrating factor is

$$e^{-\int(x^{-1}+x^{-4})dx}=x^{-1}e^{\frac{1}{3}x^{-3}} \quad (§ 13).$$

$$\therefore yx^{-1}e^{\frac{1}{3}x^{-3}}=c\int x^{-4}e^{\frac{1}{3}x^{-3}}dx+c'=-ce^{\frac{1}{3}x^{-3}}+c', \text{ or}$$

$$y+cx=c'xe^{-\frac{1}{3}x^{-3}}.$$

Ex. 7. $x^2(1-x^2)D^2y-x^3Dy-2y=0.$

Ex. 8. $x^2D^3y-5xD^2y+(4x^4+5)Dy-8x^3y=0.$

Ex. 9. $\dfrac{d^2y}{dx^2}+f(x)\dfrac{dy}{dx}+\phi(y)\left(\dfrac{dy}{dx}\right)^2=0.$*

* This type of equation was first treated by Joseph Liouville (1809–1882). Let the student show that it is the differential equation corresponding to a primitive of the form $F(y)=a\Phi(x)+b$, where F and Φ are any functions of their respective variables, and a and b are the arbitrary constants to be eliminated.

By inspection $\left(\frac{dy}{dx}\right)^{-1}$ is seen to be an integrating factor.

Introducing this, we have

$$\left(\frac{dy}{dx}\right)^{-1}\frac{d^2y}{dx^2}+f(x)+\phi(y)\frac{dy}{dx}=0;$$

whence $$\log\left(\frac{dy}{dx}\right)+\int f(x)\,dx+\int\phi(y)dy=c,$$

or $$e^{\int\phi(y)dy}dy=ce^{-\int f(x)dx}dx;$$

and $$\int e^{\int\phi(y)dy}dy=c\int e^{-\int f(x)dx}dx+c'.$$

Ex. 10. $\frac{d^2y}{dx^2}+2\cot x\frac{dy}{dx}+2\tan y\left(\frac{dy}{dx}\right)^2=0.$

61. Transformation of Variables. — In case the equation to be integrated does not come under any of the heads already treated, it is possible, at times, to reduce it to one of them by a transformation. No general rule for this can be formulated. The form of the equation will frequently suggest the transformation to be tried.

Ex. 1. $x^2y\frac{d^2y}{dx^2}+\left(x\frac{dy}{dx}-y\right)^2=0.$

The set of terms $\left(x\frac{dy}{dx}-y\right)^2$ suggests the transformation $y=vx$. Making this transformation, the equation becomes (after dropping the factor x^3)

$$xv\frac{d^2v}{dx^2}+2v\frac{dv}{dx}+x\left(\frac{dv}{dx}\right)^2=0.$$

This is exact, and has for first integral

$$xv\frac{dv}{dx}+\frac{1}{2}v^2=c.$$

This is also exact, giving $$xv^2=c_1x+c_2,$$

or $$y^2=c_1x^2+c_2x.$$

[A less obvious transformation is $y^2=v$. Let the student solve the problem by making this transformation.]

Ex. 2. $x^3\frac{d^2y}{dx^2}-\left(x\frac{dy}{dx}-y\right)^2=0.$

Ex. 3. $y\frac{d^2y}{dx^2}-\left(\frac{dy}{dx}\right)^2=y^2\log y-x^2y^2.$ [Let $\log y=v$, i.e. $y=e^v$.]

Ex. 4. $\sin^2 x\frac{d^2y}{dx^2}-2y=0.$ [Let $\cot x=z$.]

If the more obvious transformation $\sin x=z$ is made, the resulting equation can be made exact by multiplying by a proper power of z, and can then be integrated.

62. Summary. — The number of classes of differential equations of higher order than the first for which a general method of solution is known is very small. We can tell by inspection

1° when the dependent variable is absent; let the lowest ordered derivative* that appears be a new variable (§ 57);

2° when the independent variable is absent; let the first derivative of the dependent variable be a new variable, and consider the

* Provided this is not also the highest ordered derivative that appears. If such is the case, let the next lower ordered derivative be a new variable.

dependent variable as the independent one (§ 58); in particular, if the equation has the form $\frac{d^2y}{dx^2}=f(y)$, this method leads to the obvious integrating factor $\frac{dy}{dx}dx$ (§ 58, Remark).

3° If the equation is linear, and a particular integral y_1 can be found when the right-hand member is made zero, let $y=y_1v$, and in the transformed equation put $\frac{dv}{dx}=p$ (§ 59).

4° If the equation is linear and of the second order, the methods of Chapter VIII may apply (§ 55).

If none of the above cases occur, test the equation for exactness (§ 60). Should this not prove to be the case, some special device must be resorted to, such as finding an integrating factor (§ 60), or finding some suitable transformation (§ 61).

As a final resort, the method of integrating in series may be tried (§ 74).

Ex. 1. $\frac{d^2y}{dx^2}=\left(\frac{dy}{dx}\right)^2+1.$

Ex. 2. $(1-x^2)\frac{d^2y}{dx^2}-x\frac{dy}{dx}=2.$

Ex. 3. $\frac{d^2y}{dx^2}+y\frac{dy}{dx}=0.$

Ex. 4. $(1+x^3)\frac{d^3y}{dx^3}+9x^2\frac{d^2y}{dx^2}+18x\frac{dy}{dx}+6y=0.$

Ex. 5. $(x^2-x)\frac{d^2y}{dx^2}+(4x+2)\frac{dy}{dx}+2y=0.$

Ex. 6. $y(1-\log y)\frac{d^2y}{dx^2}+(1+\log y)\left(\frac{dy}{dx}\right)^2=0.$

Ex. 7. $\frac{d^2y}{dx^2}+\frac{1}{x}\frac{dy}{dx}=0.$

Ex. 8. $x(x+2y)\frac{d^2y}{dx^2}+2x\left(\frac{dy}{dx}\right)^2+4(x+y)\frac{dy}{dx}+2y+x^2=0.$

Ex. 9. $\frac{d^2y}{dx^2}+\left(\frac{dy}{dx}\right)^2+1=0.$

Ex. 10. $(1-x^2)\frac{d^2y}{dx^2}-\frac{1}{x}\frac{dy}{dx}+x^2=0.$

Ex. 11. $4x^2\frac{d^3y}{dx^3}+8x\frac{d^2y}{dx^2}+\frac{dy}{dx}=0.$

Ex. 12. $\sin x\frac{d^2y}{dx^2}-\cos x\frac{dy}{dx}+2\sin x\cdot y=0.$

Ex. 13. Determine the curves in which the radius of curvature is equal to the normal, (*a*) when the two have the same direction, (*b*) when they have opposite directions.

The radius of curvature $=\dfrac{\pm\left[1+\left(\frac{dy}{dx}\right)^2\right]^{\frac{3}{2}}}{\frac{d^2y}{dx^2}}$. The normal being supposed drawn toward the axis of x, when it and the radius of curvature are drawn in the same direction, y and $\frac{d^2y}{dx^2}$ have opposite signs; and when drawn in opposite directions, y and $\frac{d^2y}{dx^2}$ have the same sign.

Ex. 14. Determine the curves in which the radius of curvature is twice the normal, (*a*) when the two have the same direction, (*b*) when they have opposite directions.

Ex. 15. Find the curves whose radius of curvature is k times the cube of the normal.

Ex. 16. A particle which sets off from a point of the axis of x, at a distance a from the origin, moves uniformly in a direction parallel to the axis of y. It is pursued by a particle which sets off at the same time from the origin, and travels with a velocity which is n times that of the former. Required the path of the latter.

[This path is usually referred to as the *curve of pursuit*. Its differential equation may be obtained from the following considerations: Let (x, y) be the coördinates of the pursuing point, (ξ, η) those of the point pursued. The path of the latter being known, we have given (1) $f(\xi, \eta) = 0$. Since the point pursued is always in the tangent to the curve of pursuit, we have (2) $\eta - y = \frac{dy}{dx}(\xi - x)$. (1) and (2) determine ξ and η in terms of $x, y, \frac{dy}{dx}$. If the velocities of the point pursued and pursuing point are as 1 : n, we have

$$n\sqrt{d\xi^2 + d\eta^2} = \sqrt{dx^2 + dy^2};$$

or taking x as the independent variable,

$$n\sqrt{\left(\frac{d\xi}{dx}\right)^2 + \left(\frac{d\eta}{dx}\right)^2} = \sqrt{1 + \left(\frac{dy}{dx}\right)^2}.$$

Substituting in this the values of ξ and η from (1) and (2), we obtain the differential equation of the curve of pursuit.]

Ex. 17. Find the velocity of the weighted end of a simple pendulum of length l, swinging in a vacuum, if at the time $t = 0, v = 0$ and $\theta = \alpha$, where α is not so small that $\sin\alpha$ may be replaced by α as a first approximation. (See Ex. 17, § 52.)

Ex. **18.** A particle moves in a straight line attracted by a force varying inversely as the square of the distance. [Equation of motion is $\dfrac{d^2x}{dt^2}=-\dfrac{k^2}{x^2}$.] If it starts with zero velocity at a distance a from the center of the force,

(*a*) find its velocity at any point in its path,

(*b*) find the time required to reach that point,

(*c*) how far will it have to move in order to acquire the same velocity with which it would arrive at the point a if it had started to move from infinity with zero initial velocity.

(*d*) Since gravity acts according to the above law, find the velocity with which a body (a meteorite, for example) will strike the surface of the earth if it falls from a distance h above the surface.

[Acceleration due to gravity at the earth's surface is usually designated by g. Hence $k^2=gR^2$, if R is the radius of the earth.]

CHAPTER X

SYSTEMS OF SIMULTANEOUS EQUATIONS

63. General Method of Solution. — It is proved in the general theory of ordinary differential equations that a system of n equations involving n dependent variables can, in general, be solved (§ 70).

We shall consider here the case of $n = 2$, the method admitting of being extended to any number. Let the equations be

$$(1) \qquad f_1[(x)_m, \ (y)_r, t] = 0,$$

$$(2) \qquad f_2[(x)_{m+p}, (y)_s, t] = 0,$$

where the highest ordered derivatives of x appearing in (1) and (2) are respectively m and $m+p$, those of y are r and s, and t is the independent variable. Of course p may be zero.

Differentiating (1) p times, we get successively

$$(3) \qquad f_3[(x)_{m+1}, (y)_{r+1}, t] = 0,$$

$$(4) \qquad f_4[(x)_{m+2}, (y)_{r+2}, t] = 0,$$

$$\cdot \quad \cdot \quad \cdot \quad \cdot \quad \cdot$$

$$(p+2) \qquad f_{p+2}[(x)_{m+p}, (y)_{r+p}, t] = 0.$$

We now have $p+2$ equations from which to eliminate x and all of its $m+p$ derivatives. In general (unless $m = 0$) this will not be sufficient, for we must have one more equation at our disposal than the number of quantities to be eliminated. We proceed now to differentiate both (2) and $(p+2)$. Since this introduces two new equations and only one new derivative of x, we see that, by repeating the process the proper number of times, the number of equations will exceed that of the quantities to be eliminated by unity. Performing the elimination, we have a single equation in y. Integrating this and substituting the value of y in (1), we have an equation in x only, which must then be solved.

Remark. — It is almost needless to add that we may first eliminate y and its derivatives, and then solve for x.

Or, we may solve for x and for y separately. In this case the constants of integration arising are not all independent. The relations among them can be found by substituting in one of the equations (1) and (2).

64. Systems of Linear Equations with Constant Coefficients.—This method can be carried out very readily in case the equations are linear and the coefficients constants. Thus consider the example

$$\begin{cases} \dfrac{dx}{dt} - \dfrac{dy}{dt} + x = \cos t. \\[2ex] \dfrac{d^2x}{dt^2} - \dfrac{dy}{dt} + 3x - y = e^{2t}. \end{cases}$$

These may be written

(1) $$(D+1)x - Dy = \cos t,$$

(2) $$(D^2+3)x - (D+1)y = e^{2t}.$$

Differentiating (1), we have

(3) $$(D^2+D)x - D^2y = -\sin t.$$

We must eliminate x, Dx, D^2x; this requires four equations. Hence we must differentiate (2) and (3). This gives rise to

(4) $$(D^3+3D)x - (D^2+D)y = 2e^{2t},$$

(5) $$(D^3+D^2)x - D^3y = -\cos t.$$

We have now five equations from which we can eliminate the four quantities x, Dx, D^2x, D^3x. By taking $-3 \times (1)$, $1 \times (2)$, $1 \times (4)$, $-1 \times (5)$, and adding, we get

$$(6) \qquad (D^3 - D^2 + D - 1)\, y = 3\, e^{2t} - 2 \cos t,$$

which is a linear equation in y only, and can be solved readily.

Before doing so, however, we shall see how (6) can be gotten directly from (1) and (2). Since, when looked upon algebraically, (3) and (5) are respectively D and D^2 times (1), and (4) is D times (2) (temporarily supposing their right-hand members to be zero), the above method of elimination amounted to subtracting $(D^2 + 3)$ times (1) from $(D + 1)$ times (2). But this is precisely the method we would have pursued in eliminating x from (1) and (2) had D and its powers been algebraic quantities instead of operators. Now, so long as the equations are linear with constant coefficients, this process is always allowable, since it involves only the operations of addition, subtraction, and multiplication with the operator D. Hence we need only write our equations in the form of (1) and (2), solve them as algebraic equations, remembering, however, that D is an operator in case there are any terms in the right-hand members. In practice it is frequently convenient to use determinants. Thus solving (1) and (2) for y, we have

$$\begin{vmatrix} D + 1 & -D \\ D^2 + 3 & -(D + 1) \end{vmatrix} y = \begin{vmatrix} D + 1 & \cos t \\ D^2 + 3 & e^{2t} \end{vmatrix}$$

or

$$(6) \qquad (D^3 - D^2 + D - 1)y = 3\, e^{2t} - 2 \cos t.$$

The complementary function is $Y = c_1 e^t + c_2 \sin t + c_3 \cos t$.

For the particular integral, try $U = ae^{2t} + bt \sin t + ct \cos t$.

Substituting this for y in (6), we find that $a = \frac{3}{5}$, $b = \frac{1}{2}$, $c = \frac{1}{2}$.

$$(7) \quad \therefore\ y = c_1 e^t + c_2 \sin t + c_3 \cos t + \frac{3}{5} e^{2t} + \frac{1}{2} t \sin t + \frac{1}{2} t \cos t.$$

To find x we may substitute this value in either (1) or (2), and solve the resulting equation in x.*

Or we can treat x exactly as we did y, that is, solve (1) and (2) directly for x. Doing this we have

$$\begin{vmatrix} D+1 & -D \\ D^2+3 & -(D+1) \end{vmatrix} x = \begin{vmatrix} \cos t & -D \\ e^{2t} & -(D+1) \end{vmatrix} = \begin{vmatrix} D & \cos t \\ D+1 & e^{2t} \end{vmatrix},$$

or

$$(8) \qquad (D^3 - D^2 + D - 1)\,x = 2\,e^{2t} + \sin t - \cos t.\dagger$$

Solving this, we have

$$(9) \qquad x = c_1'\,e^t + c_2' \sin t + c_3' \cos t + \frac{2}{5}e^{2t} + \frac{1}{2}t \cos t.$$

But these constants are not independent of those in (7). They may be found by substituting (7) and (9) in either of the original equations and equating coefficients. Doing this, we find $c_1' = \frac{1}{2}c_1$, $c_2' = \frac{1}{2}(c_2 - c_3) + \frac{3}{4}$, $c_3' = \frac{1}{2}(c_2 + c_3) + \frac{1}{4}$.

* In general, in solving for the variable first eliminated, it is necessary to solve a differential equation. For example, if we put the value of y given by (7) in (1), we have an equation of the first order to solve; if we put it in (2), we have an equation of the second order to solve. The new constants of integration that arise now are not arbitrary, but must be determined so that the other equation is also satisfied. This is done by substituting in the other equation, and equating coefficients. In this particular example it would have been simpler to have solved for x first. The value of y could then be gotten immediately from the equation resulting from subtracting (1) from (2). Let the student do this.

† We see by this method of solution that the differential equations in x and in y, each resulting from the elimination of the other variable, have the same left-hand members, and that the complementary functions are therefore of the same form in the case of the two variables. This is obviously true in the case of n dependent variables defined by n linear equations with constant coefficients.

Hence the general solution of our system of equations is

$$4x = 2c_1e^t + (2c_2 - 2c_3 + 3)\sin t + (2c_2 + 2c_3 + 1)\cos t + \frac{8}{5}e^{2t} + 2t\cos t,$$

$$y = c_1e^t + c_2\sin t + c_3\cos t + \frac{3}{5}e^{2t} + \frac{1}{2}t(\sin t + \cos t).$$

Ex. 1. $\begin{cases} 3\dfrac{dx}{dt} + 3x + 2y = e^t, \\ 4x - 3\dfrac{dy}{dt} + 3y = 3t. \end{cases}$

Ex. 2. $\begin{cases} 2\dfrac{d^2y}{dt^2} - \dfrac{dx}{dt} - 4y = 2t, \\ 4\dfrac{dx}{dt} + 2\dfrac{dy}{dt} - 3x = 0. \end{cases}$

Ex. 3. $\begin{cases} \dfrac{d^2x}{dt^2} - 3x - 4y = 0, \\ \dfrac{d^2y}{dt^2} + x + y = 0. \end{cases}$

65. Systems of Equations of the First Order. — If the equations are of the first order, we can suppose them solved for the first derivatives of each of the dependent variables. [We shall consider the case of two dependent variables. But the methods here brought out obviously apply to the case of n such variables.] Let the system be

$$(1) \qquad \begin{cases} \dfrac{dx}{dt} = \dfrac{P(x, y, t)}{R(x, y, t)}, \\ \dfrac{dy}{dt} = \dfrac{Q(x, y, t)}{R(x, y, t)}. \end{cases}$$

The general method of § 63 applies. But in certain cases the solution can be brought about very much more readily. It is some of these cases that we shall consider now. (1) can be written in the more symmetrical form

(2) $$\frac{dx}{P}=\frac{dy}{Q}=\frac{dt}{R}.$$

1° One of these equations may involve only two of the variables, or it may be possible, by a proper choice of a pair of members of (2), to strike out a common factor so as to obtain an equation involving only two of the variables. Thus, to fix the ideas, suppose that t does not appear in $\frac{dx}{P}=\frac{dy}{Q}$, or can be removed from it. We have, on solving this,

(3) $$\phi(x, y)=c_1.$$

If this can be done a second time, so that a second relation of the form

(4) $$\psi(y, t)=c_2$$

can be found, then the complete solution consists of the two relations (3) and (4).

Ex. 1. $\frac{dx}{yt}=\frac{dy}{tx}=\frac{dt}{xy}.$

From the first two members we have $x^2-y^2=c_1$.
From the last two members we have $y^2-t^2=c_2$.

(Using the first and last members we get $t^2-x^2=c$. But this is obviously not distinct from the other two.)

2° If we can find only one integral expression of the above type, say (3), we can, by means of it, express one of the variables in terms of c_1, and the other; thus, to fix the ideas, we can solve (3) for x in terms of c_1 and y. Substituting this value of x in $\frac{dy}{Q}=\frac{dt}{R}$, we have an equation involving y, t, and the constant c_1. Solving this, we have a second relation

(5) $$\psi(y, t, c_1) = c_2.$$

(3) and (5) together constitute the general solution.

At times it is desirable to replace c_1 in (5) by its value in terms of x and y. The solution is then

$$\begin{cases} \phi(x, y) = c_1, \\ \psi(y, t, \phi) = c_2. \end{cases}$$

Ex. 2. $\dfrac{dx}{xt} = \dfrac{dy}{yt} = \dfrac{dt}{xy}$.

Here we have $\dfrac{dx}{x} = \dfrac{dy}{y}$, whence $\dfrac{x}{y} = c_1$.

$\therefore x = c_1 y$; and we have

$$\frac{dy}{yt} = \frac{dt}{c_1 y^2}, \text{ or } c_1 y\, dy = t\, dt,$$

whence $$c_1 y^2 - t^2 = c_2,$$

or $$xy - t^2 = c_2.$$

$\therefore$ The solution is $\begin{cases} x - c_1 y = 0, \\ xy - t^2 = c_2. \end{cases}$

3° It sometimes happens that we can find multipliers $\lambda(x, y, t)$, $\mu(x, y, t)$, $\nu(x, y, t)$ such that, making use of the fact that

$$\frac{dx}{P} = \frac{dy}{Q} = \frac{dt}{R} = \frac{\lambda\, dx + \mu\, dy + \nu\, dt}{\lambda P + \mu Q + \nu R},$$

(*a*) this last member when combined with one of the others gives rise to an equation which can be solved; or (*b*) $\lambda P + \mu Q + \nu R$ may equal zero, at the same time that $\lambda\, dx + \mu\, dy + \nu\, dt = 0$ satisfies the condition for integrability (§ 35); or

(*c*) by a choice of two sets of multipliers,

$$\frac{\lambda_1\, dx + \mu_1\, dy + \nu_1\, dt}{\lambda_1 P + \mu_1 Q + \nu_1 R} = \frac{\lambda_2\, dx + \mu_2\, dy + \nu_2\, dt}{\lambda_2 P + \mu_2 Q + \nu_2 R}$$

may be solvable.

If we can find two independent relations by any of these methods, each involving an arbitrary constant, we have the general solution.

Ex. 3. $\dfrac{dx}{y} = \dfrac{dy}{x} = \dfrac{dt}{t}.$

From $\dfrac{dx}{y} = \dfrac{dy}{x}$ we have $x^2 - y^2 = c_1.$

Letting $\lambda = \mu = 1$, $\nu = 0$, we have

$$\frac{dx + dy}{x + y} = \frac{dt}{t} \text{ whence } x + y = c_2 t.$$

Ex. 4. $\dfrac{dx}{cy - bt} = \dfrac{dy}{at - cx} = \dfrac{dt}{bx - ay}.$

Letting $\lambda = a$, $\mu = b$, $\nu = c$, we have by composition that the common ratio is equal to $\dfrac{a\, dx + b\, dy + c\, dt}{0}.$

$$\therefore a\, dx + b\, dy + c\, dt = 0, \text{ whence } ax + by + ct = c_1.$$

Similarly, letting $\lambda = x$, $\mu = y$, $\nu = t$, we have

$$x\, dx + y\, dy + t\, dt = 0, \text{ whence } x^2 + y^2 + t^2 = c_2.$$

Ex. 5. $\dfrac{dx}{x^2 + y^2} = \dfrac{dy}{2\,xy} = \dfrac{dt}{(x + y)\,t}.$

Letting $\lambda_1 = 1$, $\mu_1 = 1$, $\nu_1 = 0$, and $\lambda_2 = 1$, $\mu_2 = -1$, $\nu_2 = 0$, we have

$$\frac{dx+dy}{(x+y)^2} = \frac{dx-dy}{(x-y)^2}, \text{ whence } \frac{1}{x+y} = \frac{1}{x-y} - c_1,$$

or

$$2y = c_1(x^2 - y^2).$$

Again $\frac{dx+dy}{(x+y)^2} = \frac{dt}{(x+y)t}$, whence $x + y = c_2 t$.

Ex. 6. $\frac{x\,dx}{yt} = \frac{y\,dy}{xt} = \frac{dt}{y}$.

Ex. 7. $\frac{dx}{-y} = \frac{dy}{x} = \frac{dt}{1+t^2}$.

Ex. 8. $\frac{dx}{yt} = \frac{dy}{xt} = \frac{dt}{x+y}$.

Ex. 9. $\frac{dx}{x^2 - y^2 - t^2} = \frac{dy}{2xy} = \frac{dt}{2xt}$.

Ex. 10. $\frac{dx}{y+t} = \frac{dy}{t+x} = \frac{dt}{x+y}$.

Ex. 11. $\frac{dx}{x^2 + y^2 + yt} = \frac{dy}{x^2 + y^2 - xt} = \frac{dt}{(x+y)t}$.

Ex. 12. $\frac{dx}{y^3 x - 2x^4} = \frac{dy}{2y^4 - x^3 y} = \frac{dt}{9t(x^3 - y^3)}$.

66. Geometrical Interpretation. — $P(x, y, z)$,* $Q(x, y, z)$, $R(x, y, z)$ may be looked upon as determining the line $\frac{X-x}{P} = \frac{Y-y}{Q} = \frac{Z-z}{R}$ through the point (x, y, z), which is any point in space. An integral curve of the system $\frac{dx}{P} = \frac{dy}{Q} = \frac{dz}{R}$ will be, then, a curve such that, at

* From now on we shall use the three letters x, y, z, instead of x, y, t.

each point of it, it is tangent to the line at that point determined by P, Q, R. The general solution, we have seen, consists of two relations, $\left\{\begin{matrix} u(x, y, z) = c_1 \\ v(x, y, z) = c_2 \end{matrix}\right\}$, involving two arbitrary constants. That is, the general solution represents a doubly infinite system of curves, which are the intersections of two singly infinite systems of surfaces.* Thus in the case of Ex. 3, § 65, the integral curves are the intersections of the family of cylinders $x^2 - y^2 = c_1$ with the family of planes $x + y - c_2 z = 0$. In § 40 we saw that a solution of the total differential equation $P\,dx + Q\,dy + R\,dz = 0$ represents a surface such that, at each point (x, y, z) of it, it is tangent to the plane $P(X - x) + Q(Y - y) + R(Z - z) = 0$, the direction cosines of whose normal are proportional to P, Q, R. Hence we see that the integral curves of $\frac{dx}{P} = \frac{dy}{Q} = \frac{dz}{R}$ cut orthogonally any integral surface of $P\,dx + Q\,dy + R\,dz = 0$. Since we can find a family of integral surfaces of $P\,dx + Q\,dy + R\,dz = 0$ only when P, Q, R satisfy the condition for integrability [§ 35, (3)], we see that only in this case will there exist a family of surfaces of which the family of integral curves of $\frac{dx}{P} = \frac{dy}{Q} = \frac{dz}{R}$ are orthogonal trajectories. Thus, since $yz\,dx + zx\,dy + xy\,dz = 0$ has $xyz = c$ as its general solution, we see from Ex. 1, § 65, that the curves of intersection of the cylinders $x^2 - y^2 = c_1$ and $y^2 - z^2 = c_2$ are cut orthogonally by the family of surfaces $xyz = c$. On the other hand, since $xz\,dx + yz\,dy + xy\,dz = 0$ does not satisfy

* Supposing P, Q, R single-valued functions of x, y, z, there passes through any point (x_0, y_0, z_0) the single curve $\left\{\begin{matrix} u(x, y, z) = u(x_0, y_0, z_0) \\ v(x, y, z) = v(x_0, y_0, z_0) \end{matrix}\right\}$, since the differential equations determine a single direction at each point in space. If P, Q, R are not all single-valued functions, that is, if the differential equations are not both of the first degree, then more than one line (or direction) will correspond to a set of values of (x, y, z), and there will be more than one integral curve passing through a point. In this case, u and v will not be single-valued, that is, when the solutions are cleared of fractions and rationalized, the constants of integration do not enter to the first degree. This is analogous to what we found in the case of a single equation of the first order in two variables (§ 20).

the condition for integrability, there is no family of surfaces which is cut orthogonally by the curves whose equations are $\left\{\begin{matrix} x - c_1 y = 0 \\ xy - z^2 = c_2 \end{matrix}\right\}$ (Ex. 2, § 65). The converse problem of finding the orthogonal trajectories of a family of surfaces whose equation is $f(x, y, z) = c$ is always possible, at least theoretically. For this necessitates solving the system

$$\frac{dx}{\frac{\partial f}{\partial x}} = \frac{dy}{\frac{\partial f}{\partial y}} = \frac{dz}{\frac{\partial f}{\partial z}}.$$

Ex. Find the orthogonal trajectories of the family of surfaces $xy = cz$.

67. Systems of Total Differential Equations. — If we have two total differential equations in three variables,*

$$(1) \qquad \left\{\begin{matrix} P_1\,dx + Q_1\,dy + R_1\,dz = 0, \\ P_2\,dx + Q_2\,dy + R_2\,dz = 0, \end{matrix}\right.$$

it can be proved (but the limits of this book will not permit our doing so here), that the general solution consists of two relations among the variables, involving two arbitrary constants. In actual practice we proceed as follows:

If each of equations (1) separately satisfies the condition for integrability [§ 35, (3)], we solve each one, and thus obtain the solution of our system.

If only one of the equations satisfies the condition for integrability, we integrate that one, obtaining a relation $\phi(x, y, z) = c_1$. Solving this for one of the variables, we replace this and its derivative in the other equation by their values, thus giving rise to an equation in two variables only. Its solution, together with $\phi(x, y, z) = c_1$, already found, constitutes the solution of the system of equations.

* The substance of this paragraph is at once applicable to the case of n equations in $n + 1$ variables.

If neither of the equations is separately integrable, it is sometimes desirable to put (1) in the form

$$(2) \qquad \frac{dx}{P} = \frac{dy}{Q} = \frac{dz}{R},$$

where $P = Q_1R_2 - Q_2R_1$, $Q = R_1P_2 - R_2P_1$, $R = P_1Q_2 - P_2Q_1$.

The methods of § 65 may now be tried. If they do not work, then taking one of the variables as the independent one, say z, the equations may be written

$$(3) \qquad \begin{cases} \dfrac{dx}{dz} = \dfrac{P}{R}, \\ \dfrac{dy}{dz} = \dfrac{Q}{R}. \end{cases}$$

The general method of § 63 applies here.

68. Differential Equations of Higher Order than the First reducible to Systems of Equations of the First Order. — Given a single equation with one dependent variable. We may suppose it solved for the highest ordered derivative; thus, suppose we have the equation

$$(1) \qquad \frac{d^3y}{dx^3} = f\left(x, y, \frac{dy}{dx}, \frac{d^2y}{dx^2}\right).$$

If we put $\qquad \dfrac{dy}{dx} = y_1$, and $\dfrac{d^2y}{dx^2} = \dfrac{dy_1}{dx} = y_2$,

(1) may be replaced by the system of three equations of the first order

$$(2) \qquad \begin{cases} \dfrac{dy}{dx} = y_1, \\ \dfrac{dy_1}{dx} = y_2, \\ \dfrac{dy_2}{dx} = f(x, y, y_1, y_2). \end{cases}$$

As an illustration, $\qquad \dfrac{d^2y}{dx^2} + y = 0$

is equivalent to the system

$$\left.\begin{aligned}\frac{dy}{dx} &= y_1\\ \frac{dy_1}{dx} &= -y\end{aligned}\right\}, \text{ or } \frac{dx}{1}=\frac{dy}{y_1}=\frac{dy_1}{-y}.$$

Using the last two terms, we have $y^2+y_1^2=c_1^2$,

$$\text{or } y_1=\sqrt{c_1^2-y^2}.$$

$$\therefore \qquad \frac{dy}{\sqrt{c_1^2-y^2}}=dx, \text{ or } \sin^{-1}\frac{y}{c_1}=x+c_2,$$

whence

$$y=c_1\sin(x+c_2).$$

In an entirely analogous manner, a system of n equations of any order in n dependent variables may be replaced by a system of equations of the first order by letting each of the derivatives of the dependent variables up to the next to the highest ordered, in the case of each variable, be a new variable. Thus, by letting $\frac{dx}{dt}=x_1$, the system of equations of § 64 may be written

$$\begin{cases}\dfrac{dx}{dt}=x_1,\\[2mm] \dfrac{dy}{dt}=x_1+x-\cos t,\\[2mm] \dfrac{dx_1}{dt}=x_1-2x+y+e^{2t}-\cos t,\end{cases}$$

or

$$\frac{dx}{x_1}=\frac{dy}{x_1+x-\cos t}=\frac{dx_1}{x_1-2x+y+e^{2t}-\cos t}=\frac{dt}{1}.$$

69. Summary. — To solve a system of n ordinary differential equations involving n dependent variables we differentiate these equations a sufficient number of times to enable us to eliminate $n-1$ of the dependent variables and all their derivatives, thus giving rise to a single equation involving only the remaining dependent variable. We integrate this, and substituting for this variable and its derivatives their values in terms of the independent variable in any $n-1$ of the equations, we have a new system of $n-1$ equations in $n-1$ dependent variables. Repeating this, we find the value of a second variable and reduce the number of equations again, and so on. Or we can treat all of the dependent variables symmetrically by solving for each one separately, and then finding the relations among the constants of integration by substituting in some one of the original equations (§ 63).

While this method is frequently not practicable, it can be carried out very readily in case the equations are linear with constant coefficients (§ 64).

If the equations are of the first order, special methods can at times be resorted to (§ 65).

A system of n total differential equations in $n+1$ variables can be written as a system of ordinary differential equations to which the methods of § 63 and § 65 apply (§ 67).

A single differential equation in one dependent variable of higher order than the first, also a system of n such in n dependent variables may be replaced by a corresponding system of differential equations of the first order, to which at times the special methods of § 65 apply (§ 68).

It is almost needless to add that if each of a system of equations involves a single dependent variable, each is to be integrated independently of the others. Thus, see examples 1, 2, 3, 4 below.

Ex. **1.** Find the path traced out by a particle moving in a vacuum and acted upon by gravity only, if it is given an initial velocity v_0 in a direction making an angle α with the horizontal plane.

Ex. **2.** If the particle in Ex. 1 moves in a medium which exerts a resisting force proportional to the velocity, find its path.

Ex. **3.** A particle moves about a center of attraction varying directly as the distance; determine its motion, if it starts to move from a point on the axis of x at a distance a from the center, and with an initial velocity v_0 making an angle α with the axis of x.

[If the attracting force is P, and r is the distance of the particle from the the center of the force, the equations of motion are

$$\frac{d^2x}{dt^2} = -P\frac{x}{r},$$

$$\frac{d^2y}{dt^2} = -P\frac{y}{r}.$$

In this case $P = k^2r$.]

Ex. 4. If the force is a repulsive one, study the motion of the particle in Ex. 3.

Ex. 5. A solid of revolution with one point of its axis of symmetry fixed, is acted upon by gravity only. Find its angular velocity and the position of the instantaneous axis of rotation in the body.

[If A, B, C are the moments of inertia of the body with respect to the principal axes of the momental ellipsoid about the fixed point, and p, q, r are the components of the angular velocity on those axes at any instant, Euler's equations are,

$$A\frac{dp}{dt} + (C - A)\, qr = 0,$$

$$A\frac{dq}{dt} - (C - A)\, rp = 0,$$

$$C\frac{dr}{dt} \qquad = 0,$$

since $B = A$.]

Ex. 6. The component of the velocity of a particle parallel to each of the coördinate axes is proportional to the product of the other two coördinates. Find its path, and the time of describing a given portion in case the curve passes through the origin.

CHAPTER XI

INTEGRATION IN SERIES

70. The Existence Theorem. — The number of classes of differential equations that can be integrated by quadratures or other purely elementary means is very small, compared with the number of possible classes of equations. In the General Theory of Differential Equations it is proved that every ordinary differential equation with one dependent variable (and every system of n equations with n dependent variables) has a solution, in general, involving a definite number of arbitrary constants. A proper understanding of the proof of this theorem implies a knowledge of the Theory of Functions, which is not assumed here. A demonstration of the theorem will be found in almost any book dealing with the subject, presupposing a knowledge of at least the elements of the Theory of Functions.*

1° For an equation of the first order $\frac{dy}{dx} = F(x,y)$,† the theorem of existence of an integral is :

* Cauchy (1789–1857) was the first to prove this theorem. In fact he gave two proofs of it, which have become classic. For a demonstration of this theorem a student familiar with the elements of the Theory of Functions may consult among other books, Murray, *Differential Equations*, p. 190; Schlesinger, *Differentialgleichungen*, Chapter I; Picard, *Traité d'Analyse*, Vol. II, Chapter XI. More recently Picard (1856–) gave another proof, which may be found in his *Traité d'Analyse*, Vol. II, p. 301, and Vol. III, p. 88, and also in the Bulletin of the New York Mathematical Society, Vol. I, pp. 12–16.

† A differential equation of the first order $f\left(x, y, \frac{dy}{dx}\right) = 0$ may be supposed solved for $\frac{dy}{dx}$, so that it takes the form $\frac{dy}{dx} = F(x, y)$.

If $F(x,y)$ is finite, continuous, and single-valued, and has a finite partial derivative with respect to y* (see Picard, Vol. II, p. 292), *as long as x and y are restricted to certain regions, then if x_0 and y_0 are a pair of values lying in these regions, we can find one integral y, and only one, which will take the value y_0 when x takes the value x_0.*

In the proof of the theorem, y is found in the form of an infinite series

$$y_0 + c_1(x - x_0) + c_2(x - x_0)^2 + \cdots + c_n(x - x_0)^n + \cdots,$$

which series satisfies the equation when substituted in it for y, and besides is convergent for values of x sufficiently near to x_0. By the change of variable $\bar{x} = x - x_0$, the differential equation takes the form

$$\frac{dy}{d\bar{x}} = \bar{F}(\bar{x}, y),$$

and the solution takes the form

$$y = y_0 + c_1 \bar{x} + c_2 \bar{x}^2 + \cdots + c_n \bar{x}^n + \cdots.$$

Since y_0 may be chosen arbitrarily (within certain limits, however), we see that in the case of a differential equation of the first order, one arbitrary constant enters.

Remark. — The existence theorem gives a sufficient condition for an integral, and moreover, it gives a form in which the integral may be put. But this condition is not always necessary. Equations for which the conditions of the theorem are not fulfilled may have integrals. In general, but not necessarily always, such integrals will then not be developable by Taylor's theorem, or they will not be unique. A few simple examples will illustrate this:

$\dfrac{dy}{dx} = \dfrac{y}{x}$, where $\dfrac{y}{x}$ becomes indeterminate for $x = 0$, $y = 0$, has the solution $y = cx$.

* Single-valued is used in the broad sense here. Although $F(x, y)$ may have several values for a single pair of values of x and y, it will be said to be *single-valued when x and y are restricted to certain regions* if, having selected some one of its possible values for a pair of values of x and y in their respective regions, it will take a definite value for every pair of values of x and y in their regions.

Thus, while $F \equiv \pm \sqrt{x + y}$ has two values for every pair of values of x and y, if we select the value $+\sqrt{2}$ for $x = 1$, $y = 1$, F will have a definite value so long as x and y are restricted to regions where both are positive.

Here y takes the value 0 for $x=0$. It is expressed as a (finite) Taylor series, but it will be noted that c is undetermined when we put $y=0$ for $x=0$; that is, unlike the cases coming under the existence theorem, there is an indefinite number of solutions satisfying the initial condition. Moreover, it should be noted that it is impossible to find a finite value for c that will enable us to assign a value to y other than 0 for $x=0$.

Again, $\frac{dy}{dx}=\frac{x+y}{x}$, where $\frac{x+y}{x}$ becomes indeterminate for $x=0, y=0$, has the solution $y=x\log x+cx$. Here y takes the value 0 for $x=0$. But it is not possible to express the integral in the form of a Taylor series in powers of x. In this case also we have an indefinite number of integrals for the one initial value 0, and no integrals for any other initial value of y.

$2\frac{dy}{dx}=\frac{1+2x}{y}$ has for solution $y=\sqrt{c+x+x^2}$. $\frac{1+2x}{y}=\infty$, when $x=0$, $y=0$. In order that $y=0$ for $x=0$, we must have $c=0$. We have then the single solution $y=\sqrt{x+x^2}$. This is not developable by Taylor's theorem in powers of x however, although it may be developed in powers of $\sqrt{x}$.

2° If we have a system of n equations of the first order involving n dependent variables, we may suppose them solved for the derivatives of each of these variables:

$$\frac{dy}{dx}=f_1(x, y, z, \cdots, w),$$

$$\frac{dz}{dx}=f_2(x, y, z, \cdots, w),$$

$$\cdot \quad \cdot \quad \cdot \quad \cdot \quad \cdot$$

$$\frac{dw}{dx}=f_n(x, y, z, \cdots, w).$$

The general existence theorem says, in this case, that *if* $f_1, f_2, \cdots f_n$ *are all regular,* as long as* $x, y, z, \cdots, w$ *remain in certain regions, then if* $x_0, y_0, z_0, \cdots, w_0$ *are in these regions, a single set of functions* $y, z, \cdots, w$ *can be found to satisfy the system of equations and to take the values* $y_0, z_0, \cdots, w_0$ *respectively when* x *takes the value* x_0.

* For definition of regular see the second footnote, p. 203.

In the proof of the theorem these are found in the form of the infinite series

$$y = y_0 + a_1(x - x_0) + a_2(x - x_0)^2 + \cdots + a_n(x - x_0)^n + \cdots,$$
$$z = z_0 + b_1(x - x_0) + b_2(x - x_0)^2 + \cdots + b_n(x - x_0)^n + \cdots,$$
$$\cdot \quad \cdot \quad \cdot \quad \cdot \quad \cdot \quad \cdot \quad \cdot \quad \cdot \quad \cdot \quad \cdot \quad \cdot$$
$$w = w_0 + k_1(x - x_0) + k_2(x - x_0)^2 + \cdots + k_n(x - x_0)^n + \cdots,$$

which series are convergent so long as x is sufficiently close to x_0.

As before, we can make the transformation $\bar{x} = x - x_0$, which will then give our series as power series in $\bar{x}$. Here $y_0, z_0, \cdots, w_0$ may be chosen arbitrarily (within certain limits). Hence we see that a system of n equations of the first order with n dependent variables, has its general solution involving n arbitrary constants. We saw (§ 68) that an equation, with one dependent variable, of the nth order may be replaced by a system of n equations of the first order, hence it follows at once that *the general solution of a differential equation of the nth order involves n arbitrary constants.* In the way in which the equivalent system of equations of the first order is found, we see that we may choose for these arbitrary constants the values which the dependent variable and each of its derivatives up to the $(n-1)$st take for a given value of the independent variable. Hence these values are, in general, at our disposal in any given problem.

Geometrically, this means:

Of the single infinity of integral curves of a differential equation of the first order and degree, a single curve passes through a given point.

Of the double infinity of integral curves of a differential equation of the second order and first degree, a single curve passes through a given point, in a given direction.

Of the triple infinity of integral curves of a differential equation of the third order and first degree, a single curve passes through a given point, in a given direction, and having a given curvature at that point.

71. Singular Solutions. — In the existence theorem of the previous section stress should be laid upon the fact that the existence of an integral of $\frac{dy}{dx}=F(x, y)$ is assured only as long as $F(x, y)$ is finite, continuous, and single-valued in the region of (x_0, y_0). If, now, our equation is given in the form

$$f(x, y, y')=0, \text{ where } y'=\frac{dy}{dx},$$

we know that, in general, y' is expressible as a finite, continuous, and single-valued function of x and y in the region of (x_0, y_0), and takes a perfectly definite finite value y'_0 for $x=x_0$, $y=y_0$. It can be shown* that this will be true as long as

$$\frac{\partial f(x, y, y')}{\partial y'}\neq 0.$$

But if

$$\frac{\partial f}{\partial y'}=0,$$

then the expression for y' in terms of x and y ceases to be single-valued in the region of (x_0, y_0). So that in the region of such values for x and y the existence theorem does not assert the existence of a solution. As a matter of fact, a solution does not exist there in general. For from

$$f=0, \frac{\partial f}{\partial y'}=0$$

we can solve for y and y', thus

$$y=\phi(x), y'=\phi_1(x);$$

and only in exceptional cases will

$$\phi_1(x)=\frac{d\phi(x)}{dx}.$$

* A proof of this theorem will be found in many works on Analysis; for example see Liebmann, *Lehrbuch der Differentialgleichungen*, p. 8.

If it should happen that $\phi_1(x)=\frac{d\phi(x)}{dx}$, then $y=\phi(x)$ is a solution of the equation; and since it is usually distinct from the general solution, it is a singular solution. Moreover, it is identical with the singular solution we encountered in Chapter V.

Similarly in the case of differential equations of higher order than the first, singular solutions may occur. Thus if there is a solution of $f(x,y,y',\cdots,y^{(n)})=0$ for which $\frac{\partial f}{\partial y^{(n)}}$ also vanishes, this solution is, in general, a singular one.*

More generally, a system of equations of the first order

$$f_1\left(x,y,z,\cdots w,\frac{dy}{dx},\cdots,\frac{dw}{dx}\right)=0,$$

$$\cdot\quad\cdot\quad\cdot\quad\cdot\quad\cdot\quad\cdot$$

$$f_n\left(x,y,z,\cdots w,\frac{dy}{dx},\cdots,\frac{dw}{dx}\right)=0,$$

(to which any system of m equations in m dependent variables is always reducible, § 68) may have singular solutions under certain conditions. See Picard, Vol. III, p. 52.

72. Integration, in Series, of an Equation of the First Order.— If the equation

$$\frac{dy}{dx}=F(x,y) \tag{1}$$

cannot be integrated by any of the known elementary methods, the existence theorem tells us that if $F(x,y)$ is finite, continuous, and single-valued in the regions containing $x=0$, $y=c_0$ (there is no loss in assuming $x=0$; since this amounts to presupposing the substitution $\bar{x}=x-x_0$ to have been made, in case $x_0\neq 0$), one and only one solution exists which takes the value of c_0 for $x=0$. But this solu-

* Liebmann, *loc. cit.* p. 113; Boole, *Differential Equations*, p. 229.

tion is given by the existence theorem in the form of an infinite series. In actual practice we assume

(2) $$y = \sum_{n=0}^{\infty} c_n x^n = c_0 + c_1 x + c_2 x^2 + \cdots + c_n x^n + \cdots,$$

and substitute this in the differential equation (1). Then equating coefficients, we may calculate as many of the c's in (2) as we please. The existence theorem vouches for the convergence of the series (2). Three cases may arise:

1° No general law of the coefficients in (2) shows itself; in this case we can only approximate the solution in actual practice.*

2° A general law of the coefficients in (2) appears; then we can write down the general term of the series, which is equivalent to saying that the whole series is known.

3° All the terms after a certain one may turn out to be zero; in this case we have the solution in finite form.

Remark. — Cases 2° and 3° seldom occur except when the equation can be solved directly. So that this method of integrating equations of the first order is not of great practical importance for the mere purpose of integrating. But for theoretical purposes it is of the greatest importance. It may be noted that while every linear equation of the first order can be solved by quadratures, it is not always possible to perform these in terms of simple functions. In such cases this method, or that of § 74, will apply at times.

Ex. 1. $\dfrac{dy}{dx} = x + y^2.$

Here $x + y^2$ is finite, continuous, and single-valued for all values of x and y.

Put $y = c_0 + c_1 x + c_2 x^2 + \cdots + c_n x^n + \cdots.$

* While the general existence theorem tells us that the series so obtained is convergent, as long as we restrict ourselves to proper values of the variables, the convergence may be slow for certain values so that the degree of approximation, even after having calculated a fairly large number of coefficients, may not be great. This will usually be true for values of the variables near any for which $F(x, y)$ ceases to be finite, continuous, or single-valued.

Substituting, we must have

$$c_1 + 2\,c_2x + 3\,c_3x^2 + \cdots + n\,c_nx^{n-1} + \cdots \equiv x + (c_0 + c_1x + \cdots)^2.$$

Equating coefficients, we have

$$c_1 = c_0^2 \qquad \therefore c_1 = \qquad c_0^2,$$

$$2\,c_2 = 2\,c_0c_1 + 1 \qquad \therefore c_2 = \frac{1}{2} \qquad + c_0^3,$$

$$3\,c_3 = 2\,c_0c_2 + c_1^2 \qquad \therefore c_3 = \frac{1}{3}c_0 \quad + c_0^4,$$

$$4\,c_4 = 2\,c_0c_3 + 2\,c_1c_2 \qquad \therefore c_4 = \frac{5}{12}c_0^2 + c_0^5,$$

$$\cdot \quad \cdot \quad \cdot \quad \cdot \quad \cdot \quad \cdot \quad \cdot \quad \cdot \quad \cdot$$

$$2\,k\,c_{2k} = 2\,c_0c_{2k-1} + 2\,c_1c_{2k-2} + \cdots + 2\,c_{k-1}c_k.$$

$$(2\,k+1)\,c_{2k+1} = 2\,c_0c_{2k} + 2\,c_1c_{2k-1} + \cdots + 2\,c_{k-1}\,c_{k+1} + c_k^2.$$

$$\cdot \quad \cdot \quad \cdot \quad \cdot \quad \cdot \quad \cdot \quad \cdot \quad \cdot \quad \cdot$$

Here each coefficient can be calculated in terms of the preceding ones, and consequently in terms of the single one c_0. We can calculate as many as we please, but no general law shows itself. Our solution

$$y = c_0 + c_0^2x + \left(\frac{1}{2} + c_0^3\right)x^2 + \left(\frac{1}{3}\,c_0 + c_0^4\right)x^3 + \cdots$$

extended as far as we care to calculate the coefficients, is only an approximation of the solution.

Ex. 2. $\dfrac{dy}{dx} = \dfrac{2\,y}{1 - x^2}.$

Here $\dfrac{2\,y}{1-x^2}$ is finite, continuous, and single-valued in the region of $(0, c_0)$, where c_0 is any value of y.

Put $y = c_0 + c_1x + c_2x^2 + \cdots + c_nx^n + \cdots$, and substitute in the equation, after having cleared of fractions. We must have

$$(1 - x^2)(c_1 + 2\,c_2x + 3\,c_3x^2 + \cdots + nc_n\,x^{n-1} + \cdots) \equiv$$
$$2(c_0 + c_1x + c_2x^2 \cdots + c_n\,x^n + \cdots).$$

Equating coefficients,

$$c_1 = 2\,c_0.$$

$$2\,c_2 = 2\,c_1 = 4\,c_0, \qquad \therefore\; c_2 = 2\,c_0.$$

$$3\,c_3 - c_1 = 2\,c_2 = 4\,c_0, \qquad \therefore\; c_3 = 2\,c_0.$$

The general law seems to be $c_n = 2\,c_0$. We shall prove this to be the case by showing that if it holds for c_n it holds for c_{n+1}. For,

$$(n+1)c_{n+1} - (n-1)c_{n-1} = 2\,c_n.$$

Now $$c_{n-1} = c_n = 2\,c_0.$$

$$\therefore\; (n+1)c_{n+1} = 2(n-1)c_0 + 4\,c_0 = 2(n+1)c_0,$$

or $$c_{n+1} = 2\,c_0.$$

Hence the solution, in the form of an infinite series, is

$$y = c_0(1 + 2\,x + 2\,x^2 + 2\,x^3 + \cdots + 2\,x^n + \cdots).$$

[This equation can be integrated by the methods of Chapter II. Let the student do this, and compare the result with the one here obtained.]

When the equation $\dfrac{dy}{dx} = F(x, y)$ is such that the various derivatives of F can be readily calculated numerically for special values of x and y, the following method is sometimes found practicable:

We have seen that the solution given by the general existence theorem has the form

$$y = c_0 + c_1(x - x_0) + c_2(x - x_0)^2 + \cdots + c_n(x - x_0)^n + \cdots.$$

Using the general form of the Taylor development of a function, we see that when

$$c_0 = y_0,\ c_1 = \left(\frac{dy}{dx}\right)_0,\ \cdots,\ c_n = \frac{1}{n\,!}\left(\frac{d^ny}{dx^n}\right)_0,\ \cdots,$$

we obtain that solution which takes the value y_0 for $x = x_0$.

From the differential equation, we have

$$c_1 = \left(\frac{dy}{dx}\right)_0 = F(x_0, y_0).$$

Differentiating the differential equation we have

$$\frac{d^2y}{dx^2} = \frac{\partial F}{\partial x} + \frac{\partial F}{\partial y}\frac{dy}{dx},$$

whence $$c_2 = \frac{1}{2\,!}\left(\frac{d^2y}{dx^2}\right)_0 = \frac{1}{2\,!}\left(\frac{\partial F}{\partial x}\right)_0 + \frac{1}{2\,!}c_1\left(\frac{\partial F}{\partial y}\right)_0.$$

Differentiating again, we find c_3 in terms of c_1 and c_2. And so on. The student will find on applying this method to the examples above that it works very readily in the case of Ex. 1.

73. Riccati's Equation. — The equation studied by Count Riccati (1676–1754), and to which his name has been given, is of the form

$$\frac{dy}{dx} + by^2 = cx^m,$$

where b, c, m are constants.* The equation in Ex. 1, § 72 is of this type. For certain special values of b, c, m this equation can be inte-

* Frequently the equation $x\dfrac{dy}{dx} - \alpha y + \beta y^2 = \gamma x^n$ is taken as the type of a Riccati equation. This is obviously reducible to the other by the transformation $z = x^a$, $y = uz$.

grated in finite terms. (See Ex. 4, § 18; also Forsyth, p. 170; Boole, Chapter VI; Johnson, Chapter IX.) But in general, the only way to get the solution is to integrate in series.

Riccati equations frequently arise, and it is often desirable to make use of the properties of their solutions without actually knowing the latter. On the other hand, it is sometimes possible to find the general solution by quadratures or by merely algebraic processes, when certain information is at hand. The following properties will at times prove of value:

We shall consider a more general form, which is now usually considered as the type of a Riccati equation,

$$\frac{dy}{dx} = X_0 + X_1 y + X_2 y^2,$$

where X_0, X_1, X_2 are functions of x or constants.

1° If a particular integral y_1 is known, the substitution $y = \frac{1}{z} + y_1$ transforms the equation to $\frac{dz}{dx} + (X_1 + 2 y_1 X_2)z = - X_2$, which is linear, and can therefore be solved by two quadratures (§ 13). Hence we have, *if a particular integral* y_1 *is known, the transformation* $y = \frac{1}{z} + y_1$ *gives rise to a linear equation in* z *which can be solved by two quadratures.*

2° Since the form of the solution of a linear equation of the first order is $z = \gamma(x) + C\delta(x)$, that is, the constant of integration enters linearly, we have that

$$y = \frac{1}{z} + y_1(x) = \frac{1}{\gamma(x) + C\delta(x)} + y_1(x)$$

$$= \frac{\alpha(x) + C\beta(x)}{\gamma(x) + C\delta(x)}.$$

Hence, *the constant of integration enters bilinearly in the general solution of the Riccati equation.*

3° The equation $y=\dfrac{\alpha+\beta C}{\gamma+\delta C}$ may be looked upon as a bilinear transformation of C into y, which latter is a particular solution as soon as a value of C is fixed. Corresponding to any four values of C, say C_1, C_2, C_3, C_4, we have y_1, y_2, y_3, y_4. Since double ratios are left unaltered by bilinear transformations, we have

$$\{y_1, y_2, y_3, y_4\}=\{C_1, C_2, C_3, C_4\}=\text{a constant.}$$

Hence, *if y_1, y_2, y_3, y_4 are any four particular integrals, the function $\dfrac{(y_4-y_3)(y_2-y_1)}{(y_4-y_1)(y_2-y_3)}$ is equal to a constant for all values of x.*

4° As a direct consequence of 3° it follows that *if we know three particular integrals y_1, y_2, y_3, the general solution is given at once by*

$$\frac{y-y_3}{y-y_1}=c\,\frac{y_2-y_3}{y_2-y_1};$$

whence $y=\dfrac{y_3(y_2-y_1)-cy_1(y_2-y_3)}{y_2-y_1-c(y_2-y_3)}$, *i.e. y is given by purely algebraic means.*

5° If y_1 and y_2 are two known particular integrals, put $z=\dfrac{y-y_1}{y-y_2}$ and take the derivative of the logarithm of both sides. This gives

$$\frac{1}{z}\frac{dz}{dx}=\frac{1}{y-y_1}\left(\frac{dy}{dx}-\frac{dy_1}{dx}\right)-\frac{1}{y-y_2}\left(\frac{dy}{dx}-\frac{dy_2}{dx}\right).$$

Since
$$\frac{dy}{dx}=X_0+X_1y+X_2y^2,$$

$$\frac{dy_1}{dx}=X_0+X_1y_1+X_2y_1^2,$$

$$\frac{dy_2}{dx}=X_0+X_1y_2+X_2y_2^2,$$

we have $$\frac{1}{z}\frac{dz}{dx} = X_2(y_1 - y_2);$$

whence $$z = Ce^{\int X_2(y_1 - y_2)dx},$$

i.e. $$\frac{y - y_1}{y - y_2} = Ce^{\int X_2(y_1 - y_2)dx},$$

from which y can be gotten at once. Hence, *if two particular integrals y_1 and y_2 are known, the transformation $z = \frac{y - y_1}{y - y_2}$ leads to an equation in which the variables are separated, so that it can be solved by a single quadrature.*

Properties 1°, 4°, 5° call attention to the close analogy of the Riccati equation to linear differential equations; for the knowledge of each additional particular integral brings us nearer to the general solution. Thus, *the knowledge of a single particular integral enables us to reduce the problem of solving a Riccati equation to one of solving a linear equation of the first order, that is to one involving two quadratures; the knowledge of two particular integrals enables us to find the general solution by performing a single quadrature; while a knowledge of three particular integrals gives us the general solution by a very simple algebraic process.*

6° As a matter of fact, *the substitution*

$$y = -\frac{1}{X_2 z}\frac{dz}{dx}$$

transforms the Riccati equation into the homogeneous linear equation of the second order

$$X_2\frac{d^2z}{dx^2} - (X_1X_2 + X'_2)\frac{dz}{dx} + X_0X_2^2z = 0,$$

where $X'_2 = \frac{dX_2}{dx}$.

[Let the student show that, conversely, the substitution $y = e^{\int v dx}$ transforms a homogeneous linear equation of the second order into a Riccati equation.]

74. Integration, in Series, of Equations of Higher Order than the First. — If the equation when solved for the highest ordered derivative*

$$\frac{d^3y}{dx^3} = F\left(x, y, \frac{dy}{dx}, \frac{d^2y}{dx^2}\right)$$

is such that the successive derivatives of F can be readily calculated numerically for special values of

$$x, y, \frac{dy}{dx}, \frac{d^2y}{dx^2},$$

a method analogous to that given at the end of § 72 may be employed.

The solution given by the general existence theorem being in the form

$$y = c_0 + c_1(x - x_0) + c_2(x - x_0)^2 + \cdots + c_n(x - x_0)^n + \cdots,$$

we know from the general form of the Taylor development of a function that

$$c_0 = y_0, \ c_1 = \left(\frac{dy}{dx}\right)_0, \ c_2 = \frac{1}{2!}\left(\frac{d^2y}{dx^2}\right)_0, \ \cdots, \ c_n = \frac{1}{n!}\left(\frac{d^ny}{dx^n}\right)_0, \cdots.$$

Now the general solution involves three arbitrary constants, and we saw (§ 70, at the end) that a particular solution will be determined as soon as we fix the values of $y, \frac{dy}{dx}, \frac{d^2y}{dx^2}$ for $x = x_0$; let us call them y_0, y_0', y_0''.

* To fix the ideas we shall illustrate with an equation of the third order, although this method, when practicable, applies to equations of any order.

From the differential equation we have at once

$$c_3 = \frac{1}{3!}\left(\frac{d^3y}{dx^3}\right)_0 = \frac{1}{3!}F(x_0, y_0, y_0', y_0'').$$

Differentiating the equation we can calculate $\left(\frac{d^4y}{dx^4}\right)_0$ in terms of

$$x_0, y_0, \left(\frac{dy}{dx}\right)_0, \left(\frac{d^2y}{dx^2}\right)_0, \left(\frac{d^3y}{dx^3}\right)_0;$$

so that we can find c_4 in terms of c_0, c_1, c_2, c_3. And so on.

When this method does not work readily, we may employ the first method of § 72 as there given. But again, only in comparatively few cases will this turn out to be practicable. The following modification of this method has been found of special service in the case of linear differential equations in which the coefficients are polynomials in x, and such that when we substitute x^m for y in the left-hand member, there results only a small number of distinct powers of x (preferably, not more than two). In the case of a Cauchy equation (§ 51) there results a single power of x, and the equation can be solved by purely algebraic means.*

If we take the equation

$$\frac{d^2y}{dx^2} - xy = 0,$$

the result of putting $y = x^m$ in the left-hand member is

$$m(m-1)x^{m-2} - x^{m+1}.$$

* Thus, putting $y = x^m$ in the left hand member of (1), § 51, we get

$$[k_0 m(m-1)\cdots(m-n+1) + k_1 m(m-1)\cdots(m-n+2) + \cdots + k_{n-1}m + k_n]x^m.$$

Equating the coefficient of x^m to zero and solving for m, we get in general n distinct particular solutions and therefore the complementary function. It is readily seen to be the same as that found in § 51. The cases of equal and complex values of m can be treated entirely analogously to those for linear equations with constant coefficients.

In some respects, the Cauchy equation is simpler than the corresponding linear equation with constant coefficients. But for actually obtaining its solution, especially when there is a right-hand member, it is usually simpler to transform it to a linear equation with constant coefficients, as was done in § 51.

Here there are two distinct powers of x which differ by 3, and $m-2$ is the smaller exponent. Hence, if we let

$$y = c_0x^m + c_1x^{m+3} + c_2x^{m+6} + \cdots + c_rx^{m+3r} + \cdots$$

we shall have a solution if

1° m is so chosen that $m(m-1)=0$, *i.e.* $m=0$ or 1,

2° the c's are so chosen that the rest of the terms cancel each other in pairs, *i.e.* we must have

$$(m+3)(m+2)c_1 - c_0 = 0,$$

$$(m+6)(m+5)c_2 - c_1 = 0,$$

$$\cdot \quad \cdot \quad \cdot \quad \cdot \quad \cdot \quad \cdot \quad \cdot \quad \cdot$$

$$(m+3r)(m+3r-1)c_r - c_{r-1} = 0,$$

$$\cdot \quad \cdot \quad \cdot \quad \cdot \quad \cdot \quad \cdot \quad \cdot \quad \cdot$$

$$\therefore c_r = \frac{1}{(m+3r)(m+3r-1)}c_{r-1}.$$

For $m=0$, we have $c_1 = \frac{1}{3\cdot 2}c_0 \quad = \frac{1}{3\,!}c_0,$

$$c_2 = \frac{1}{6\cdot 5}c_1 \quad = \frac{1\cdot 4}{6\,!}c_0,$$

$$c_3 = \frac{1}{9\cdot 8}c_2 \quad = \frac{1\cdot 4\cdot 7}{9\,!}c_0,$$

$$c_4 = \frac{1}{12\cdot 11}c_3 = \frac{1\cdot 4\cdot 7\cdot 10}{12\,!}c_0,$$

$$\cdot \quad \cdot \quad \cdot \quad \cdot \quad \cdot \quad \cdot$$

$$c_r = \qquad = \frac{1\cdot 4\cdot 7\cdots[1+3(r-1)]}{(3r)\,!}c_0$$

$$\cdot \quad \cdot \quad \cdot \quad \cdot \quad \cdot \quad \cdot \quad \cdot \quad \cdot$$

$$\therefore c_0\left(1+\frac{1}{3!}x^3+\frac{1\cdot 4}{6!}x^6+\frac{1\cdot 4\cdot 7}{9!}x^9+\cdots\right.$$
$$\left.+\frac{1\cdot 4\cdot 7\cdots[1+3(r-1)]}{(3r)!}x^{3r}+\cdots\right)$$

is an integral. Let us call it Ay_1, where A, like c_0, is an arbitrary constant.

For $m=1$ we have $c_1=\frac{1}{4\cdot 3}c_0 = \frac{2}{4!}c_0,$

$$c_2=\frac{1}{7\cdot 6}c_1 = \frac{2\cdot 5}{7!}c_0,$$

$$c =\frac{1}{10\cdot 9}c_2=\frac{2\cdot 5\cdot 8}{10!}c_0,$$

. . . . . .

$$c_r= \quad =\frac{2\cdot 5\cdot 8\cdots[2+3(r-1)]}{(1+3r)!}c_0.$$

. . . .

$$\therefore c_0x\left(1+\frac{2}{4!}x^3+\frac{2\cdot 5}{7!}x^6+\cdots+\frac{2\cdot 5\cdot 8\cdots[2+3(r-1)]}{(1+3r)!}x^{3r}+\cdots\right)$$

is also an integral. Let us call it By_2, where B is an arbitrary constant. y_1 and y_2 are obviously linearly independent. Moreover, they are convergent by the general existence theorem. Hence, $y=Ay_1+By_2$ is the general solution, since it contains two arbitrary constants.

If the right-hand member of the equation had not been zero, we would have proceeded to find the particular integral by a similar method. Thus, suppose the equation had been

$$\frac{d^2y}{dx^2}-xy=2x^{-3}.$$

Since the result of putting $y = c_0x^m$ in the left-hand member is

$$c_0m(m-1)x^{m-2} - c_0x^{m+1},$$

$c_0x^m + c_1x^{m+3} + \cdots + c_rx^{m+3r} + \cdots$ will be a particular integral, provided $c_0m(m-1)x^{m-2} = 2x^{-3}$, and the other terms that arise destroy each other in pairs.

The first of these will be true if

$$m - 2 = -3, \qquad \therefore m = -1,$$

and

$$c_0m(m-1) = 2, \qquad \therefore c_0 = 1.$$

The other terms will destroy each other in pairs if

$$c_1 = \frac{1}{2 \cdot 1}c_0 \qquad \therefore c_1 = \frac{1}{2\,!}$$

$$c_2 = \frac{1}{5 \cdot 4}c_1 \qquad \therefore c_2 = \frac{3}{5\,!}$$

$$c_3 = \frac{1}{8 \cdot 7}c_2 \qquad \therefore c_3 = \frac{3 \cdot 6}{8\,!}$$

. . . .

$$c_r = \frac{1}{(3r-1)(3r-2)}c_{r-1} \quad \therefore c_r = \frac{3 \cdot 6 \cdot 9 \cdots 3(r-1)}{(3r-1)\,!}.$$

. . . .

Hence a particular integral is

$$x^{-1} + \frac{1}{2\,!}x^2 + \frac{3}{5\,!}x^5 + \cdots + \frac{3 \cdot 6 \cdot 9 \cdots 3(r-1)}{(3r-1)\,!}x^{3r-1} + \cdots$$

The above example suggests a general method,* in case the result of substituting $y = x^m$ in the left-hand member of the equation gives

* As mentioned in the Remark, § 72, this method applies to linear equations of the first order as well as to those of higher orders.

rise to only two* distinct powers of x, say $f(m)x^h + \phi(m)x^{h+l}$, where l is a positive integer.

By the manner in which $f(m)$ and $\phi(m)$ arise, it is obvious that one of them at least must be of degree n in m.

I. To find the complementary function we proceed as follows:

(a) Suppose $f(m)$ is of degree n in m, so that $f(m) = 0$ has n roots, $m_1, m_2, m_3, \cdots m_n$, all distinct, or some repeated.

Letting $y = c_0x^m + c_1x^{m+l} + \cdots + c_rx^{m+rl} + \cdots$, we have, on substituting in the left-hand member of the differential equation,

$$\begin{aligned}
&c_0f(m)x^h + c_0\phi(m)x^{h+l}\\
&+ c_1f(m+l)x^{h+l} + c_1\phi(m+l)x^{h+2l}\\
&+ c_2f(m+2l)x^{h+2l} + c_2\phi(m+2l)x^{h+3l}\\
&+ \cdot \quad \cdot \quad \cdot \quad \cdot \quad \cdot \quad \cdot \quad \cdot \quad \cdot\\
&+ c_{r-1}f(m+[r-1]l)x^{h+(r-1)l} + c_{r-1}\phi(m+[r-1]l)x^{h+rl}\\
&+ c_rf(m+rl)x^{h+rl} + c_r\phi(m+rl)x^{h+(r+1)l}\\
&+ \cdot \quad \cdot \quad \cdot \quad \cdot \quad \cdot \quad \cdot \quad \cdot \quad \cdot \quad \cdot \quad \cdot
\end{aligned}$$

This will be zero provided

1° $f(m) = 0$, *i.e.* $m = m_1, m_2, \cdots, m_n$;

2° $c_rf(m+rl) + c_{r-1}\phi(m+[r-1]l) = 0$, for $\begin{Bmatrix} r = 1, 2, 3, \cdots, \infty \\ m = m_1, m_2, \cdots, m_n \end{Bmatrix}$,

$$\text{or } c_r = -\frac{\phi(m+[r-1]l)}{f(m+rl)}c_{r-1}$$

$$= (-1)^r\frac{\phi(m+[r-1]l)\,\phi(m+[r-2]l)\cdots\phi(m+l)\,\phi(m)}{f(m+rl)f(m+[r-1]l)\cdots f(m+2l)f(m+l)}c_0.$$

* An analogous method can be deduced in the case of a larger number, but this will not be done here.

To each value of m corresponds, then, in general, a particular integral.* If any c vanishes, all that follow do so, and that integral appears in finite form.

If two values of m are equal, of course, the same particular integral will correspond to these. Moreover, if two of the m's differ by an integral multiple of l, say $m_2 = m_1 + gl$, then corresponding to the smaller value m_1, the coefficient c_g will be infinite [since $f(m_2) = f(m_1 + gl) = 0$], unless the numerator is also zero. Hence the method here given gives us only as many particular solutions in general, as there are distinct roots of $f(m) = 0$, whose differences are not multiples of l. The remaining solutions must then be sought by a modification of our process.†

If $f(m)$ is of lower degree than n, while the above method may lead to infinite series which will satisfy the equation, the general theory gives us no assurance that they are convergent. So that, unless the solutions come out in finite form, it is best to make use of the fact that if $f(m)$ is not of the degree n, $\phi(m)$ is.

(b) If $\phi(m)$ is of degree n in m, $\phi(m) = 0$ will be satisfied by n values of m, say $m'_1, m'_2, \cdots m'_n$.

Letting $y = c_0 x^m + c_{-1} x^{m-l} + c_{-2} x^{m-2l} + \cdots + c_{-r} x^{m-rl} + \cdots$, we have on substituting in the left-hand member of the equation,

$$\begin{array}{ll} c_0\,\phi(m)x^{k+l} & + c_0 f(m)x^k \\ + c_{-1}\,\phi(m-l)x^k & + c_{-1} f(m-l)x^{k-l} \\ + c_{-2}\,\phi(m-2l)x^{k-l} & + c_{-2} f(m-2l)x^{k-2l} \\ + \cdot \quad \cdot \quad \cdot \quad \cdot & \cdot \quad \cdot \quad \cdot \quad \cdot \end{array}$$

* It should be noted that corresponding to any m which is not a positive integer, we have a solution which is not a power series, but x^m multiplied by a power series.

Although the general existence theorem no longer applies here, because the coefficients in the equation, when the leading coefficient is made unity, cease to be finite for $x = 0$, the general theory of linear equations assures us of the convergence of the series for certain values of x. (Schlesinger, *Differentialgleichungen*, § 24.)

† In this case the particular solutions, which the general method fails to give, usually involve logarithmic terms. Thus see Ex. 2.

$$+c_{-r+1}\phi(m-[r-1]\,l)x^{\lambda-(r-2)l}+c_{-r+1}f(m-[r-1]l)x^{\lambda-(r-1)l}$$
$$+c_{-r}\,\phi(m-rl)x^{\lambda-(r-1)l}+c_{-r}f(m-rl)x^{\lambda-rl}$$
$$+\ \cdot\quad\cdot\quad\cdot\quad\cdot\quad\cdot\quad\cdot\quad\cdot\quad\cdot$$

This will be zero if

1° $\phi(m)=0$, *i.e.* $m=m'_1, m'_2, \cdots, m'_n$,

$$2^\circ\quad c_{-r}=-\frac{f(m-[r-1]l)}{\phi(m-rl)}c_{-r+1}\ \text{ for }\ \begin{Bmatrix} r=1,2,3,\cdots,\infty \\ m=m'_1, m'_2,\cdots, m'_n\end{Bmatrix}$$

$$=(-1)^r\frac{f(m-[r-1]l)f(m-[r-2]l)\cdots f(m-l)f(m)}{\phi(m-rl)\phi(m-[r-1]l)\cdots\phi(m-2l)\phi(m-l)}c_0.$$

To each value of m corresponds, in general, a particular integral of the form $x^m(c_0+c_{-1}x^{-l}+c_{-2}x^{-2l}+\cdots+c_{-r}x^{-rl}+\cdots)$.

Here x^m is multiplied by a power series going according to negative ascending powers of x. The latter reduces at once to an ordinary power series in t if we put $t=\frac{1}{x}\cdot$ Of course, if any of the roots of $\phi(m)=0$ are repeated, or differ by a multiple of l, the number of integrals obtained by this method will be less than n.

It should be noted that the integrals found by methods (a) and (b) in case both $f(m)$ and $\phi(m)$ are of the nth degree, are not distinct. In general (but not necessarily), they are different in form, but only n of the functions defined by them can be linearly independent.*

* An infinite series defines a function for those values of the variable only for which the series is convergent. The series found by methods (a) and (b), being developments in the region of the origin and of ∞ respectively, may not, and frequently do not, converge for the same values of x. Hence if the series are infinite, it is usually impossible to compare them. But if the functions represented by each set can be "continued" into the region of the other, then a linear relation will be found to exist among any $n+1$ of them.

II. To find the particular integral in case the right hand member is a power of x, say Ax^s, we proceed as follows:

If $f(m)$ is of degree n, we must have, using the results found in connection with method (a) above:

1° $h = s$. Since h is a linear function of m, this will determine a single value of m, say m_s.

2° $c_0 f(m_s) = A$.

3° The remaining coefficients are determined as in (a) except that now m_s is used for m.

This method will be in default when $f(m_s) = 0$.

Or using the results found in connection with method (b)*, we get the solution by putting:

1° $h + l = s$. This will determine a value m_s' of m.

2° $c_0 \phi(m_s') = A$.

3° The remaining coefficients are determined as in (b) except that now m_s' is used for m.

If $\phi(m_s') = 0$ at the same time that $f(m_s) = 0$, the particular integral will not be of the form here sought. Other means will have to be used to find it.

For a perfectly general method for finding the particular integral see Schlesinger, § 54.

Ex. 1. $(x - x^2)\dfrac{d^2y}{dx^2} - 3\dfrac{dy}{dx} + 2y = x + 3x^4.$

Substituting x^m for y in the left-hand member, we have

$$m(m-4)x^{m-1} - [m(m-1) - 2]x^m.$$

* Whether $\phi(m)$ is of degree n or not. The objection to using method (a) or (b) in case $f(m)$ or $\phi(m)$ is not of degree n is, that if the solution comes out in the form of an infinite series, the latter need not be convergent. If the solution appears in finite form, the method applies perfectly well; if not, the convergence of the series must be looked into. In the case of the example worked out above, if the right-hand member had been x, method (a) would have given the particular integral in the form of an infinite series, which turns out to be $y_1 - 1$. On the other hand, method (b) gives -1 for the particular integral at once.

$$\therefore l = 1,$$

$$f(m) = m(m-4),$$

$$\phi(m) = -(m+1)(m-2).$$

Since $f(m)$ is of degree 2, we shall use method (a).

Then $m = 0$ or 4,

and $$c_r = -\frac{\phi(m+[r-1])}{f(m+r)}c_{r-1} = \frac{m+r-3}{m+r-4}c_{r-1}.$$

For $m = 0$ we have $$c_1 = \frac{1-3}{1-4}c_0 = \frac{2}{3}c_0,$$

$$c_2 = \frac{2-3}{2-4}c_1 = \frac{1}{3}c_0,$$

$$c_3 = \frac{3-3}{3-4}c_2 = 0,$$

$$c_4 = \qquad = 0,$$

. . . . .

$\therefore\ c_0\left(1 + \frac{2}{3}x + \frac{1}{3}x^2\right)$, or $A(3 + 2x + x^2)$ is an integral. Let us call it Ay_1.

For $m = 4$ we have $$c_1 = \frac{1+1}{1}c_0 = 2\,c_0,$$

$$c_2 = \frac{2+1}{2}c_1 = 3\,c_0,$$

$$c_3 = \frac{3+1}{3}c_2 = 4\,c_0,$$

$$c_4 = \frac{4+1}{4}c_3 = 5\,c_0,$$

. . . . .

$\therefore\ c_0(x^4 + 2x^5 + 3x^6 + \cdots + nx^{n+3} + \cdots)$ is an integral. Let us call it By_2.

Therefore $Ay_1 + By_2$ is the complementary function.

Since $\phi(m)$ is of the second degree, we can also use method (b).

Then $m = 2$ or -1,

and
$$c_{-r} = -\frac{f(m-[r-1])}{\phi(m-r)}c_{-r+1} = \frac{m-r-3}{m-r-2}c_{-r+1}.$$

For $m = 2$ we have

$$c_{-1} = \frac{-1-1}{-1}c_0 = 2c_0,$$

$$c_{-2} = \frac{-2-1}{-2}c_{-1} = 3c_0,$$

$$c_{-3} = \frac{-3-1}{-3}c_{-2} = 4c_0,$$

. . . . .

$\therefore\ c_0(x^2 + 2x + 3 + 4x^{-1} + 5x^{-2} + \cdots + nx^{3-n} + \cdots)$ is an integral. Let us call it $A'y_1'$.

For $m = -1$ we have

$$c_{-1} = \frac{-1-4}{-1-3}c_0 = \frac{5}{4}c_0,$$

$$c_{-2} = \frac{-2-4}{-2-3}c_{-1} = \frac{6}{4}c_0,$$

$$c_{-3} = \frac{-3-4}{-3-3}c_{-2} = \frac{7}{4}c_0,$$

. . . . .

$\therefore\ \frac{c_0}{4}(4x^{-1} + 5x^{-2} + 6x^{-3} + \cdots + (n+3)x^{-n} + \cdots)$ is an integral. Let us call it $B'y_2'$.

Hence $A'y_1' + B'y_2'$ is the complementary function.

Here it turns out that $y_1' - y_2' = y_1$.

To find a particular integral, we consider each term of the right-hand member separately.

For the term x we have, putting

$$c_0 m(m-4)x^{m-1} = x,$$

$$m = 2 \text{ and } c_0 = -\frac{1}{4}.$$

For $m = 2$
$$c_r = \frac{r-1}{r-2}c_{r-1}.$$

$$\therefore c_1 = 0, \text{ and all that follow are zero.}$$

Hence $-\frac{1}{4}x^2$ is a particular integral corresponding to x.

Corresponding to $3x^4$, we have, putting

$$c_0 m(m-4)x^{m-1} = 3x^4,$$

$$m = 5 \text{ and } c_0 = \frac{3}{5}.$$

For $m = 5$
$$c_r = \frac{r+2}{r+1}c_{r-1}.$$

$$\therefore c_1 = \frac{3}{2}c_0 = \frac{3}{10}\cdot 3,$$

$$c_2 = \frac{4}{3}c_1 = \frac{3}{10}\cdot 4,$$

$$c_3 = \frac{5}{4}c_2 = \frac{3}{10}\cdot 5,$$

$$c_4 = \frac{6}{5}c_3 = \frac{3}{10}\cdot 6,$$

$$\cdot \quad \cdot \quad \cdot \quad \cdot$$

Hence
$$\frac{3}{10}[2x^5 + 3x^6 + 4x^7 + \cdots + (n+1)x^{n+4} + \cdots]$$

is a particular integral corresponding to $3x^4$. Comparing this with y_2 above, we see that it is equal to $\frac{3}{10}(y_2 - x^4)$.

Hence a particular integral is $-\frac{3}{10}x^4$.*

The complete solution is then

$$y = Ay_1 + By_2 - \frac{1}{4}x^2 - \frac{3}{10}x^4.$$

Ex. 2. $x\frac{d^2y}{dx^2} + \frac{dy}{dx} + y = 0.$

The result of putting $y = x^m$ in the left-hand member is

$$m^2 x^{m-1} + x^m.$$

$$\therefore l = 1,$$

$$f(m) = m^2,\ \phi(m) = 1.$$

Hence method (*a*) applies. $m = 0, 0,$

$$c_r = -\frac{1}{f(m+r)}c_{r-1} = -\frac{1}{r^2}c_{r-1}$$

since $m = 0$ is the only choice for m.

$$\therefore c_1 = -\frac{1}{1^2}c_0 = -c_0,$$

$$c_2 = -\frac{1}{2^2}c_1 = \frac{1}{2^2}c_0,$$

$$c_3 = -\frac{1}{3^2}c_2 = -\frac{1}{(3!)^2}c_0,$$

$$\cdot\quad\cdot\quad\cdot\quad\cdot\quad\cdot$$

$$\therefore c_0\left(1 - \frac{x}{(1!)^2} + \frac{x^2}{(2!)^2} - \frac{x^3}{(3!)^2} + \cdots + (-1)^n\frac{x^n}{(n!)^2} + \cdots\right)$$

is an integral. Let us call it cy_1.

* This form could have been obtained at once by equating $c_0\phi(m)x^m$ to $3x^4$.

To find a second integral, let $y = y_1 v + w$, where v and w are two functions still to be determined. Since the single requirement of having this value of y satisfy the equation is imposed upon these two functions, a second relation between them may be assumed. We shall do this so as to simplify our work as much as possible.

Substituting this value of y in the equation, and remembering that y_1 is an integral, we get

$$x\frac{d^2w}{dx^2} + \frac{dw}{dx} + w + y_1\left(x\frac{d^2v}{dx^2} + \frac{dv}{dx}\right) + 2x\frac{dy_1}{dx}\frac{dv}{dx} = 0.$$

Assume now that $$x\frac{d^2v}{dx^2} + \frac{dv}{dx} = 0$$

(this is the second relation at our disposal). Then

$$v = A + B\log x.$$

And our equation to determine w is

$$x\frac{d^2w}{dx^2} + \frac{dw}{dx} + w = -2B\frac{dy_1}{dx}$$

$$= 2B - \frac{2Bx}{2!} + \frac{2Bx^2}{2!3!} - \frac{2Bx^3}{3!4!} + \cdots + (-1)^n\frac{2Bx^n}{n!(n+1)!} + \cdots.$$

Letting $w = c_0 + c_1 x + c_2 x^2 + \cdots + c_n x^n + \cdots,$

whence $$\frac{dw}{dx} = c_1 + 2c_2 x + 3c_3 x^2 + \cdots + (n+1)c_{n+1}x^n + \cdots,$$

$$x\frac{d^2w}{dx^2} = 2c_2 x + 6c_3 x^2 + \cdots + n(n+1)c_{n+1}x^n + \cdots,$$

we have on equating coefficients

$$c_0 + c_1 = 2B,$$

$$c_1 + 4c_2 = -\frac{2B}{2!},$$

$$c_2 + 9c_3 = \frac{2B}{2!\,3!},$$

$$\cdot \quad \cdot \quad \cdot \quad \cdot$$

$$c_n + (n+1)^2 c_{n+1} = (-1)^n \frac{2B}{n!\,(n+1)!},$$

$$\cdot \quad \cdot \quad \cdot \quad \cdot \quad \cdot \quad \cdot \quad \cdot \quad \cdot$$

Since we are looking for as simple a particular integral as possible, there will be no loss in assuming $c_0 = 0$; then

$$c_1 = 2B,$$

$$c_2 = -\frac{1}{(2!)^2}\left(1 + \frac{1}{2}\right)2B,$$

$$c_3 = \frac{1}{(3!)^2}\left(1 + \frac{1}{2} + \frac{1}{3}\right)2B.$$

$$\cdot \quad \cdot \quad \cdot \quad \cdot \quad \cdot$$

$$c_n = (-1)^{n+1}\frac{1}{(n!)^2}\left(1 + \frac{1}{2} + \frac{1}{3} + \cdots + \frac{1}{n}\right)2B.$$

$$\cdot \quad \cdot \quad \cdot \quad \cdot \quad \cdot \quad \cdot \quad \cdot \quad \cdot$$

Hence the general solution is

$$y = (A + B\log x)\left[1 - \frac{x}{(1!)^2} + \frac{x^2}{(2!)^2} - \frac{x^3}{(3!)^2} + \cdots + (-1)^n\frac{x^n}{(n!)^2} + \cdots\right]$$

$$+ 2B\left[\frac{x}{(1!)^2} - \frac{x^2}{(2!)^2}\left(1 + \frac{1}{2}\right) + \frac{x^3}{(3!)^2}\left(1 + \frac{1}{2} + \frac{1}{3}\right) + \cdots\right.$$

$$\left. + (-1)^{n+1}\frac{x^n}{(n!)^2}\left(1 + \frac{1}{2} + \frac{1}{3} + \cdots + \frac{1}{n}\right) + \cdots\right].$$

Ex. 3. $(x - x^2)\frac{d^2y}{dx^2} + 4\frac{dy}{dx} + 2y = 0.$

Ex. 4. $2x^2\frac{d^2y}{dx^2} - x\frac{dy}{dx} + (1 - x^2)y = x^2.$

Ex. 5. $(x - x^2)\frac{d^2y}{dx^2} + 3\frac{dy}{dx} + 2y = 0.$

75. Gauss's Equation. Hypergeometric Series.—The integration of the equation

$$(Az^2 + Bz + C)\frac{d^2y}{dz^2} + (Dz + E)\frac{dy}{dz} + Fy = 0,$$

where A, B, C, D, E, F are constants and $B^2 - 4AC \neq 0$, leads to a remarkable series exhaustively studied by Karl Friedrich Gauss (1777–1855). This series and its differential equation were discovered by Euler (1707–1783). Putting $z = ax + b$, we have

$$[Aa^2x^2 + (2Aab + Ba)x + Ab^2 + Bb + C]\frac{d^2y}{dx^2} + (D'x + E')\frac{dy}{dx} + F'y = 0,$$

where D', E', F' are constants.

Choosing a and b so that

$$Ab^2 + Bb + C = 0$$

and

$$2Ab + B = -Aa \neq 0,$$

and dividing by the coefficient of $(x^2 - x)\frac{d^2y}{dx^2}$, we have

$$(x^2 - x)\frac{d^2y}{dx^2} + (Px + Q)\frac{dy}{dx} + Ry = 0,$$

where P, Q, R are constants.

Substituting x^m for y in the left-hand member, we have

$$-m(m-1-Q)x^{m-1}+[m^2-(1-P)m+R]x^m.$$

Putting $Q=-\gamma$, $1-P=-(\alpha+\beta)$, $R=\alpha\beta$,

our equation takes the form

$$(x^2-x)\frac{d^2y}{dx^2}+[(\alpha+\beta+1)x-\gamma]\frac{dy}{dx}+\alpha\beta y=0,*$$

and the result of substituting $y=x^m$ in the left-hand member is

$$f(m)x^{m-1}+\phi(m)x^m,$$

where $f(m)=-m(m-1+\gamma)$, $\phi(m)=(m+\alpha)(m+\beta)$.

Using the method (*a*) of § 74, we have

$$m=0 \text{ or } 1-\gamma.$$

$$c_r=-\frac{\phi(m+r-1)}{f(m+r)}c_{r-1}=\frac{(m+r+\alpha-1)(m+r+\beta-1)}{(m+r)(m+r+\gamma-1)}c_{r-1}.$$

For $m=0$, we have

$$c_1=\qquad\qquad=\frac{\alpha\cdot\beta}{1\cdot\gamma}c_0,$$

$$c_2=\frac{(\alpha+1)(\beta+1)}{2(\gamma+1)}c_1=\frac{\alpha(\alpha+1)\beta(\beta+1)}{1\cdot2\cdot\gamma(\gamma+1)}c_0,$$

. . . .

$$c_n=\qquad\qquad=\frac{\alpha(\alpha+1)\cdots(\alpha+n-1)\beta(\beta+1)\cdots(\beta+n-1)}{1\cdot2\cdot3\cdots n\cdot\gamma(\gamma+1)\cdots(\gamma+n-1)}c_0,$$

. . . .

* This is usually referred to as *Gauss's equation.*

Putting $c_0 = 1$, we have the particular integral

$$y_1 = 1 + \frac{\alpha \cdot \beta}{1 \cdot \gamma} x + \frac{\alpha(\alpha+1)\beta(\beta+1)}{1 \cdot 2 \cdot \gamma(\gamma+1)} x^2 + \cdots$$

$$+ \frac{\alpha(\alpha+1)\cdots(\alpha+n-1)\beta(\beta+1)\cdots(\beta+n-1)}{1 \cdot 2 \cdot 3 \cdots n \cdot \gamma(\gamma+1)\cdots(\gamma+n-1)} x^n + \cdots.$$

This is the *hypergeometric series*, and is usually represented by

$$F(\alpha, \beta, \gamma, x).$$

For $m = 1 - \gamma$, the student should show that the integral is $c_0 y_2$, where $y_2 = x^{1-\gamma} F(\alpha - \gamma + 1, \beta - \gamma + 1, 2 - \gamma, x)$.

If α or β is a negative integer, while γ is not, y_1 reduces to a polynomial.

If γ is a negative integer (including zero), say $\gamma = -g \overline{\gtrless} 0$, while neither α nor β is one of the integers from $-g$ to 0, the coefficients in y_1 beginning with the gth become infinite. So that this form for y_1 cannot be used. (Another form involving $\log x$ can be found, which together with y_2 will give the general solution.)

If $\gamma = -g \overline{\gtrless} 0$, where g is an integer, and α or β is one of the integers from $-g$ to 0, y_1 reduces to a polynomial (excepting in the case when α or $\beta = \gamma$; in this case a factor in the numerator and one in the denominator of every coefficient beginning with the gth term vanish; here these zero factors neutralize each other, and the result obtained by striking out these zero factors gives us an available form for y_1.*) In this case, although the two roots of $f(m) = 0$ differ by an integer, no logarithmic term enters in the general solution.

If γ is a positive number, say $\gamma = g > 0$, and neither α nor β is one of the integers from 1 to g, the coefficients in y_2 become infinite. In this case a new form for y_2 involving $\log x$ can be found.

* See Schlesinger, *Differentialgleichungen*, § 34.

If, however, $\gamma = g > 0$ where g is an integer, and α or β is one of the integers from 1 to g, y_2 reduces to a polynomial (with the exceptional case of α or $\beta = \gamma$, which is handled in the same manner as the exceptional case for γ a negative integer or zero). So that in this case also no logarithmic term enters in the general solution.

Using the method (*b*) of § 74, let the student show that

$$y_1' = \frac{1}{x^\alpha} F\left(\alpha,\ 1+\alpha-\gamma,\ 1+\alpha-\beta,\ \frac{1}{x}\right),$$

$$y_2' = \frac{1}{x^\beta} F\left(\beta,\ 1+\beta-\gamma,\ 1+\beta-\alpha,\ \frac{1}{x}\right)$$

are a pair of linearly independent integrals.

If y_1 and y_2 are infinite series with finite coefficients, they will be convergent for all values of x less than 1 in absolute value, and divergent for all values of x greater than 1 in absolute value; while y_1' and y_2' are convergent as long as x is greater than 1 in absolute value, and divergent as long as x is less than 1 in absolute value.

The hypergeometric series may at times represent well-known functions. Thus let the student show that

Ex. 1. $F(-n, \beta, \beta, -x) = (1+x)^n$ for β any constant.

Ex. 2. $xF(1, 1, 2, -x) = \log(1+x)$.

Ex. 3. $\text{Limit}_{\beta=\infty} F\left(1, \beta, 1, \frac{x}{\beta}\right) = e^x$.

Ex. 4. $\text{Limit}_{\alpha,\beta=\infty}\, xF\left(\alpha, \beta, \frac{3}{2}, -\frac{x^2}{4\alpha\beta}\right) = \sin x$.

Ex. 5. Express as hypergeometric series the following functions:

$$\frac{1}{1-x},\ (1+x)^n + (1-x)^n,\ (1+x)^n - (1-x)^n,\ \cos x,\ e^x + e^{-x}.$$

For further examples see Gauss, Collected Works, Vol. III, p. 127

CHAPTER XII

PARTIAL DIFFERENTIAL EQUATIONS

76. Primitives involving Arbitrary Constants. — Partial differential equations may be obtained from primitives involving either arbitrary constants or arbitrary functions. Thus, consider the family of spheres of fixed radius R, with their centers lying in the plane of $z=0$. The equation of this family is obviously

$$(x-a)^2+(y-b)^2+z^2=R^2, \tag{1}$$

where a and b are arbitrary constants.

We shall consider x and y as independent variables, and shall put

$$\frac{\partial z}{\partial x}=p,\ \frac{\partial z}{\partial y}=q,\ \frac{\partial^2 z}{\partial x^2}=r,\ \frac{\partial^2 z}{\partial x \partial y}=s,\ \frac{\partial^2 z}{\partial y^2}=t.$$

Differentiating (1) with respect to x and y respectively, we get

$$x-a+zp=0, \tag{2}$$

$$y-b+zq=0. \tag{3}$$

Eliminating a and b from (1), (2), (3) we get

$$z^2(p^2+q^2+1)=R^2, \tag{4}$$

which is known as the *differential equation* corresponding to the *primitive* (1); on the other hand (1) is said to be a *solution* of (4).

Perfectly generally, if we start with any relation involving two arbitrary constants and three variables, of which two are independent,

(1) $$\phi(x, y, z, a, b) = 0,$$

where z is taken as the dependent variable, we get on differentiating with respect to x, and then to y

(2) $$\frac{\partial\phi}{\partial x} + p\frac{\partial\phi}{\partial z} = 0,$$

(3) $$\frac{\partial\phi}{\partial y} + q\frac{\partial\phi}{\partial z} = 0$$

We now have three equations from which to eliminate a and b. Doing this, there results

(4) $$f(x, y, z, p, q) = 0,$$

a differential equation of the first order which has (1) for its primitive.

If the primitive involves more than two arbitrary constants, more than three equations will be necessary to eliminate these, so that if only two independent variables are involved, the resulting differential equation will be of higher order than the first. Thus consider

(5) $$ax^2 + by^2 + cz^2 = 1.$$

Differentiating with respect to x and y respectively, we have

(6) $$ax + czp = 0,$$

(7) $$by + czq = 0.$$

We have now only three equations from which to eliminate a, b, c. Hence we must differentiate again. Differentiating (6) with respect to x we have

(8) $$a + c(zr + p^2) = 0.$$

Eliminating a, b, c from (5), (6), (7), (8), we have

$$\begin{vmatrix} x^2 & y^2 & z^2 & -1 \\ x & 0 & zp & 0 \\ 0 & y & zq & 0 \\ 1 & 0 & zr+p^2 & 0 \end{vmatrix} = 0, \text{ or}$$

(9) $$(xzr + xp^2 - zp) = 0,$$

which is a differential equation having (5) for its primitive. Had we differentiated (6) with respect to y, we would have gotten

(8′) $$czs + cpq = 0;$$

whence, on eliminating a, b, c from (5), (6), (7), (8′),

(9′) $$zs + pq = 0,$$

which is also a differential equation having (5) for its primitive. Again, differentiating (7) with respect to y, we get

(8″) $$b + c(zt + q^2) = 0.$$

Eliminating a, b, c from (5), (6), (7), (8″), we have

(9″) $$yzt + yq^2 - zq = 0,$$

which is also a differential equation having (5) for its primitive.

Since (9), (9′), (9″) are all of the second order, there is no choice among them, and all may be said to be differential equations belonging to (5)*.

* Attention should be called to the fact that there is no such ambiguity in the case of the equation of the first order belonging to a primitive involving two constants. Since there are $k+1$ kth derivatives of z depending on the number of times we differentiate with respect to x and to y, it is clear that, in order that a primitive should give rise to a unique partial differential equation of the nth order, the number of arbitrary constants it contains must be equal to the sum of all the integers from 2 to $n+1$, *i.e.* $\frac{n(n+3)}{2}$.

Form the differential equations having the following equations for primitives, a, b, c being the arbitrary constants to be eliminated:

Ex. 1. $(x-a)^2+(y-a)^2+z^2=b^2$.

Ex. 2. $a(x^2+y^2)+bz^2=1$.

Ex. 3. $z=ax+by+\sqrt{a^2+b^2}$.

Ex. 4. $z=(x+a)(y+b)$.

Ex. 5. $(x-a)^2+(y-b)^2+(z-c)^2=1$.

Ex. 6. $z=ax+by+cxy$.

Ex. 7. $ax+by+cz=1$.

77. Primitives involving Arbitrary Functions. — Let u and v be two known functions of x, y, z, and suppose we have the arbitrary relation

(1) $$\phi(u, v)=0.$$

Differentiating with respect to x and y respectively, we have

(2) $$\frac{\partial\phi}{\partial u}\left(\frac{\partial u}{\partial x}+\frac{\partial u}{\partial z}p\right)+\frac{\partial\phi}{\partial v}\left(\frac{\partial v}{\partial x}+\frac{\partial v}{\partial z}p\right)=0,$$

(3) $$\frac{\partial\phi}{\partial u}\left(\frac{\partial u}{\partial y}+\frac{\partial u}{\partial z}q\right)+\frac{\partial\phi}{\partial v}\left(\frac{\partial v}{\partial y}+\frac{\partial v}{\partial z}q\right)=0.$$

In order that (2) and (3) may hold simultaneously, we must have

$$\begin{vmatrix}\dfrac{\partial u}{\partial x}+\dfrac{\partial u}{\partial z}p & \dfrac{\partial u}{\partial y}+\dfrac{\partial u}{\partial z}q\\ \dfrac{\partial v}{\partial x}+\dfrac{\partial v}{\partial z}p & \dfrac{\partial v}{\partial y}+\dfrac{\partial v}{\partial z}q\end{vmatrix}=0, \text{ or}$$

(4) $$Pp+Qq=R,$$

where

$$P=\begin{vmatrix}\frac{\partial u}{dz} & \frac{\partial u}{\partial y}\\ \frac{\partial v}{\partial z} & \frac{\partial v}{\partial y}\end{vmatrix},\quad Q=\begin{vmatrix}\frac{\partial u}{\partial x} & \frac{\partial u}{\partial z}\\ \frac{\partial v}{\partial x} & \frac{\partial v}{\partial z}\end{vmatrix},\quad R=\begin{vmatrix}\frac{\partial u}{\partial y} & \frac{\partial u}{\partial x}\\ \frac{\partial v}{\partial y} & \frac{\partial v}{\partial x}\end{vmatrix}.$$

This is a linear partial differential equation of the first order. By *a linear partial differential equation of the first order* we mean one that is linear in the derivatives of the dependent variable (the way in which the dependent variable itself enters playing no rôle).*

If u and v are two known functions of x, y, z, and we have

$$(1)\qquad f[x, y, z, \phi(u), \psi(v)] = 0,$$

where f is some known function, but ϕ and ψ are arbitrary functions, then, on differentiating, we have

$$(2)\qquad \frac{\partial f}{\partial x}+\frac{\partial f}{\partial z}p+\frac{\partial f}{\partial \phi}\phi'(u)\left(\frac{\partial u}{\partial x}+\frac{\partial u}{\partial z}p\right)+\frac{\partial f}{\partial \psi}\psi'(v)\left(\frac{\partial v}{\partial x}+\frac{\partial v}{\partial z}p\right)=0,$$

$$(3)\qquad \frac{\partial f}{\partial y}+\frac{\partial f}{\partial z}q+\frac{\partial f}{\partial \phi}\phi'(u)\left(\frac{\partial u}{\partial y}+\frac{\partial u}{\partial z}q\right)+\frac{\partial f}{\partial \psi}\psi'(v)\left(\frac{\partial v}{\partial y}+\frac{\partial v}{\partial z}q\right)=0.$$

Since (2) and (3) introduce two new functions $\phi'(u)$, $\psi'(v)$, five equations are necessary to eliminate the four functions $\phi(u)$, $\psi(v)$, $\phi'(u)$, $\psi'(v)$.

We must differentiate (2) and (3), which will give rise to three new equations involving second derivatives of z, and the new functions $\phi''(u)$ and $\psi''(v)$. In all we have now six equations from which, in general, it is not possible to eliminate the six arbitrary functions. If such is the case, we must differentiate again, this time obtaining four new equations, involving third derivatives of z, and the two new

* The student will note the difference between this definition and that given for a linear ordinary differential equation of the first order (§ 13). See also § 87.

functions $\phi'''(u)$ and $\psi'''(v)$. That is, we have now ten equations, from which the eight functions can be eliminated, in general, in two distinct ways. Hence we are led in the general case to two differential equations of the third order.

In the case of special forms of the function f, we can eliminate the six functions $\phi(u)$, $\phi'(u)$, $\phi''(u)$, $\psi(v)$, $\psi'(v)$, $\psi''(v)$ from the first six equations arising in the above process. Thus, for example, suppose $f \equiv w-[\phi(u)+\psi(v)]=0$, where w is a known function of x, y, z. Then

$$\frac{\partial w}{\partial x}+\frac{\partial w}{\partial z}p-\left[\phi'(u)\left(\frac{\partial u}{\partial x}+\frac{\partial u}{\partial z}p\right)+\psi'(v)\left(\frac{\partial v}{\partial x}+\frac{\partial v}{\partial z}p\right)\right]=0,$$

$$\frac{\partial w}{\partial y}+\frac{\partial w}{\partial z}q-\left[\phi'(u)\left(\frac{\partial u}{\partial y}+\frac{\partial u}{\partial z}q\right)+\psi'(v)\left(\frac{\partial v}{\partial y}+\frac{\partial v}{\partial z}q\right)\right]=0.$$

These last two equations involve only $\phi'(u)$ and $\psi'(v)$, and not $\phi(u)$ and $\psi(v)$. Hence, since the three equations gotten by differentiating these involve only $\phi'(u)$, $\phi''(u)$, $\psi'(v)$, $\psi''(v)$, these four functions can be eliminated from the five equations in which they enter and a single differential equation of the second order arises which has $w=\phi(u)+\psi(v)$ for its primitive for any form of the functions ϕ and ψ.*

Find the differential equations arising from the following primitives:

Ex. 1. $\phi(x+y+z, x^2+y^2+z^2)=0$.

Letting $x+y+z=u$, $x^2+y^2+z^2=v$, we have $\phi(u, v)=0$.

* The five equations are linear in $\phi'(u)$, $\phi''(u)$, $\psi'(v)$, $\psi''(v)$. Hence the elimination can be effected readily. Moreover r, s, t enter linearly also, and in such a way that the resulting differential equation is also linear in them, that is, it has the form $Rr+Ss+Tt=V$, where R, S, T, V, are functions of x, y, z, p, q.

Differentiating, we get

$$\frac{\partial\phi}{\partial u}(1+p)+\frac{\partial\phi}{\partial v}(2x+2zp)=0,$$

$$\frac{\partial\phi}{\partial u}(1+q)+\frac{\partial\phi}{\partial v}(2y+2zq)=0.$$

$$\therefore (1+p)(y+zq)-(1+q)(x+zp)=0,$$

or

$$(y-z)p+(z-x)q=x-y.$$

Ex. 2. $\phi\left(z^2-xy, \dfrac{y}{x}\right)=0.$

Ex. 3. $\phi(x^2+y^2, z-xy)=0.$

Ex. 4. $z=\phi(x+y)+\psi(x-y).$

Ex. 5. $z=\phi(x+y)+\psi(xy).$

78. Solution of a Partial Differential Equation.—Since a primitive which gives rise to a differential equation is obviously a solution of that equation, we see that arbitrary functions as well as arbitrary constants may enter into the solutions of partial differential equations. By the general existence theorem* for partial differential equations, it is seen that every partial differential equation, or system of such equations, has a solution containing a definite number of arbitrary functions. As an arbitrary function may contain an indefinite number of arbitrary constants, a solution involving an arbitrary function is much

* This theorem is due to Cauchy, as is that for ordinary differential equations. Proofs of the theorem have also been given by Darboux and Mme. Sophie de Kowalewski, among others. The proof of the latter is the most readily followed and is the one usually given. See Goursat-Bourlet, *Équations aux Dérivées Partielles du Premier Ordre*, Chapter I; also Picard, *Traité d'Analyse*, Vol. II, p. 318.

more general than one that contains any fixed number of arbitrary constants. We shall speak of the solution given by the existence theorem (which is a solution involving one or more arbitrary functions) as *the general solution.*

The mere statement of the existence theorem for a system of partial differential equations of any order is quite complicated. We shall give here simply a statement in the cases of a single equation of the first and second orders in three variables:

1° Consider the equation $f(x, y, z, p, q) = 0$. We shall suppose that p actually appears.* Solve for it, so that the equation may be supposed to have the form

$$p = F(x, y, z, q).$$

The existence theorem tells us that *if $F(x, y, z, q)$ is regular* † *in the regions of $x = x_0$, $y = y_0$, $z = z_0$, $q = q_0$, and if $\phi(y)$ is any arbitrarily chosen function of y, regular in the region of y_0, and such that $\phi(y_0) = z_0$, $\phi'(y_0) = q_0$, there exists one and only one solution $z = \psi(x, y)$, which is regular in the regions of $x = x_0$, $y = y_0$, and which reduces to $z = \phi(y)$* for $x = x_0$.

Geometrically this means that *given any curve $z = \phi(y)$ in the plane $x = x_0$ there exists one and only one surface (in any region for which there are no singular points* ‡ *of the differential equation) passing through that curve.*

This can be generalized. By a proper choice of coördinates it can be shown that one integral surface, and only one, can be found passing through any arbitrarily chosen curve, whether plane or twisted (as long as we avoid singular points of the equation). See Goursat-Bourlet, p. 21.

2° Consider the equation $f(x, y, z, p, q, r, s, t) = 0$. If this contains neither r nor t, a linear transformation will introduce one or both of these. We shall

* If p is absent, then q must appear, and the argument here employed can be carried out by interchanging in it p and q, and x and y.

† The function $F(x, y, z, q)$ is said to be *regular* in the regions of $x = x_0$, $y = y_0$, $z = z_0$, $q = q_0$, if it can be developed by Taylor's theorem in a convergent power series in $x - x_0$, $y - y_0$, $z - z_0$, $q - q_0$.

‡ By a *singular point* of the equation we mean one whose coördinates together with the corresponding value of q determine a set of values in the regions of which F ceases to be regular.

suppose then that one of them appears; and there will be no loss in supposing that it is r. Solving for this, the equation takes the form

$$r = F(x, y, z, p, q, s, t).$$

The existence theorem tells us that *if F is regular in the regions of $x = x_0$, $y = y_0$, $z = z_0$, $p = p_0$, $q = q_0$, $s = s_0$, $t = t_0$, and if $\phi(y)$ and $\psi(y)$ are any arbitrarily chosen functions of y, regular in the region of y_0, and such that $\phi(y_0) = z_0$, $\phi'(y_0) = q_0$, $\phi''(y_0) = t_0$, $\psi(y_0) = p_0$, $\psi'(y_0) = s_0$, there exists one and only one function, z, of x and y which is regular in the regions of $x = x_0$, $y = y_0$, and such that $z = \phi(y)$ and $p = \psi(y)$ for $x = x_0$.*

Looked at geometrically, this means that *given any curve $z = \phi(y)$ in the plane $x = x_0$, there exists an indefinite number of integral surfaces passing through it. But if at each point of this curve we fix a tangent plane, there exists one and only one integral surface through the curve and tangent to these planes.* For, the direction cosines of the normal to the tangent plane are proportional to $p, q, -1$. As soon as the curve is given we know $\phi(y)$. The q at each point of this curve is $\phi'(y)$, hence it is also known. So that to give the tangent plane at each point is to give p, which is our $\psi(y)$. Once $\phi(y)$ and $\psi(y)$ are given, the existence theorem says the integral surface is determined uniquely.

As in the previous case, this may be extended to apply to any curve whether plane or twisted.

Here again, no singular points of the differential equation are supposed to appear in the regions in which we are interested.

CHAPTER XIII

PARTIAL DIFFERENTIAL EQUATIONS OF THE FIRST ORDER

79. Linear Partial Differential Equations of the First Order. Method of Lagrange. — Lagrange deduced the following very neat method of solving linear partial differential equations of the first order. The general type of such an equation for one dependent and two independent variables is

$$(1)\qquad Pp + Qq = R,$$

where P, Q, R are functions of x, y, z.

Consider the linear equation with three independent variables x, y, z,

$$(2)\qquad P\frac{\partial u}{\partial x} + Q\frac{\partial u}{\partial y} + R\frac{\partial u}{\partial z} = 0,$$

which is *homogeneous* (*i.e.* there is no right-hand member), and the coefficients are functions of the independent variables only.

If $u = c$ satisfies (1), u will be a solution of (2); for, differentiating,

we have $$p = -\frac{\frac{\partial u}{\partial x}}{\frac{\partial u}{\partial z}}, \qquad q = -\frac{\frac{\partial u}{\partial y}}{\frac{\partial u}{\partial z}}.$$

Substituting these in (1) we get (2).

Conversely, if u is a solution of (2), $u=c$ will satisfy (1). For, from (2) we have, on solving for R,

$$-\frac{\frac{\partial u}{\partial x}}{\frac{\partial u}{\partial z}}P-\frac{\frac{\partial u}{\partial y}}{\frac{\partial u}{\partial z}}Q=R.$$

But
$$-\frac{\frac{\partial u}{\partial x}}{\frac{\partial u}{\partial z}}=p,\quad -\frac{\frac{\partial u}{\partial y}}{\frac{\partial u}{\partial z}}=q,$$

when $u=c$. Hence (1) follows. From this we see that the problem of solving (1) is equivalent to that of solving (2).

Consider now the system of ordinary differential equations,

(3)
$$\frac{dx}{P}=\frac{dy}{Q}=\frac{dz}{R}.$$

If u is a solution of (2), $u=c$ satisfies (3); for, if we multiply numerator and denominator of these fractions by $\frac{\partial u}{\partial x}, \frac{\partial u}{\partial y}, \frac{\partial u}{\partial z}$ respectively, we have, by composition, that each of the fractions of (3) is equal to

$$\frac{\frac{\partial u}{\partial x}dx+\frac{\partial u}{\partial y}dy+\frac{\partial u}{\partial z}dz}{P\frac{\partial u}{\partial x}+Q\frac{\partial u}{\partial y}+R\frac{\partial u}{\partial z}}.$$

Since the denominator is zero by hypothesis, the numerator (which is du) must be. Hence $u=c$ is a solution of (3) [§ 65, 3°, (*b*)].

Conversely, if $u=c$ is a solution of (3), u will satisfy (2). For, by differentiation we have

(4)
$$\frac{\partial u}{\partial x}dx+\frac{\partial u}{\partial y}dy+\frac{\partial u}{\partial z}dz=0.$$

To say that $u=c$ satisfies (3) is to say that for it

$$dx : dy : dz = P : Q : R.$$

Hence, replacing in (4) the former by the latter, we have

$$P\frac{\partial u}{\partial x}+Q\frac{\partial u}{\partial y}+R\frac{\partial u}{\partial z}=0,$$

which shows that (2) is satisfied.

Hence the problem of solving (2) is reduced to that of solving (3).

Therefore, finally, the problem of solving (1) is reduced to that of solving (3), since every solution of the latter is also a solution of the former, and conversely.

Moreover, if $u=c_1$, $v=c_2$ are any two independent solutions of (3), any function of u and v, say $\phi(u, v)$, will satisfy (2). For, substituting this in the left-hand member of (2), we get

$$\frac{\partial \phi}{\partial u}\left(P\frac{\partial u}{\partial x}+Q\frac{\partial u}{\partial y}+R\frac{\partial u}{\partial z}\right)+\frac{\partial \phi}{\partial v}\left(P\frac{\partial v}{\partial x}+Q\frac{\partial v}{\partial y}+R\frac{\partial v}{\partial z}\right).$$

This vanishes, since u and v each satisfy (2). Hence $\phi(u, v)$ is also a solution of (2) irrespective of the choice of the function ϕ.

Therefore, $\phi(u, v)=0$ * is the general solution of (1), since it contains an arbitrary function, § 78. Since ϕ is an arbitrary function, there is no loss in putting zero for the right-hand member instead of an arbitrary constant.

Word for word, the above proof may be extended to a linear equation with n independent variables. So that we can formulate the rule:

To find the general solution of

$$P_1\frac{\partial z}{\partial x_1}+P_2\frac{\partial z}{\partial x_2}+\cdots+P_n\frac{\partial z}{\partial x_n}=R,$$

solve

$$\frac{dx_1}{P_1}=\frac{dx_2}{P_2}=\cdots=\frac{dx_n}{P_n}=\frac{dz}{R}.$$

* The solution may obviously also be written in the form $u=f(v)$, or $v=\psi(u)$.

If the general solution of this system is

$$u_1 = c_1,\ u_2 = c_2,\ \cdots,\ u_n = c_n,$$

then $\phi(u_1, u_2, \cdots, u_n) = 0$, *where* ϕ *is an arbitrary function of* $u_1, u_2, \cdots, u_n$, *will be the general solution of* (1).

Ex. 1. $xzp + yzq = xy.$

$$\frac{dx}{xz} = \frac{dy}{yz} = \frac{dz}{xy}.$$

Multiplying numerator and denominator of the three members by $y, x, -2z$ respectively, we have, by composition, since the denominator vanishes,

$$y\,dx + x\,dy - 2z\,dz = 0. \quad \text{(Method 3°, (b), § 65.)}$$

$$\therefore xy - z^2 = c_1.$$

Besides, $\dfrac{dx}{xz} = \dfrac{dy}{yz}$, or $\dfrac{dx}{x} = \dfrac{dy}{y}$ gives $\dfrac{y}{x} = c_2.$

$$\therefore \phi\left(xy - z^2,\ \frac{y}{x}\right) = 0 \text{ is the general solution.}$$

Ex. 2. $-y\dfrac{\partial u}{\partial x} + x\dfrac{\partial u}{\partial y} + (1 + z^2)\dfrac{\partial u}{\partial z} = 0.$

$$\frac{dx}{-y} = \frac{dy}{x} = \frac{dz}{1 + z^2} = \frac{du}{0}.$$

We have at once $\qquad x\,dx + y\,dy = 0,$

or $\qquad x^2 + y^2 = c_1.$

Also, $\qquad u = c_2.$

Multiplying numerator and denominator of the first two members by $-y$ and x respectively, we have by composition,

$$\frac{x\,dy - y\,dx}{x^2+y^2} = \frac{dz}{1+z^2}. \quad \text{(Method 3°, (a), § 65.)}$$

$$\therefore \tan^{-1}\frac{y}{x} - \tan^{-1} z = c,$$

or

$$\frac{y-xz}{x+yz} = c_3.$$

$\therefore \phi\left(u, x^2+y^2, \frac{y-xz}{x+yz}\right) = 0$, or $u = f\left(x^2+y^2, \frac{y-xz}{x+yz}\right)$ is the general solution.

Ex. 3. $yp - xq = x^2 - y^2$.

Ex. 4. $(y-z)p + (z-x)q = x - y$.

80. Integrating Factors of the Ordinary Differential Equation $M\,dx + N\,dy = 0$. — We are now in a position to treat satisfactorily the problem of finding an integrating factor for an ordinary differential equation of the first order and degree. We have seen (§ 7) that the necessary and sufficient condition that μ be an integrating factor of $M\,dx + N\,dy = 0$ is

$$\frac{\partial(\mu N)}{\partial x} - \frac{\partial(\mu M)}{\partial y} = 0,$$

or

$$\mu\left(\frac{\partial N}{\partial x} - \frac{\partial M}{\partial y}\right) + N\frac{\partial \mu}{\partial x} - M\frac{\partial \mu}{\partial y} = 0, \text{ whence}$$

(1)

$$\frac{N}{\frac{\partial M}{\partial y} - \frac{\partial N}{\partial x}}\frac{\partial \mu}{\partial x} + \frac{M}{\frac{\partial N}{\partial x} - \frac{\partial M}{\partial y}}\frac{\partial \mu}{\partial y} = \mu.*$$

* Since the number of solutions of this linear partial differential equation is infinite, we see again that an ordinary differential equation of the first order has an infinite number of integrating factors (§ 5).

To find μ satisfying (1) we consider the system of ordinary equations,

$$(2)\qquad \frac{\frac{\partial M}{\partial y}-\frac{\partial N}{\partial x}}{N}\,dx=\frac{\frac{\partial N}{\partial x}-\frac{\partial M}{\partial y}}{M}\,dy=\frac{d\mu}{\mu}.$$

Remark.—In actual practice, when trying to integrate $M\,dx+N\,dy=0$, we are not desirous of finding the most general form for μ; as a matter of fact, as a rule, the simpler the form the better. Hence any one solution of (2) will be sufficient. It should be noted that this is not usually a practical method. But, for special forms of M and N, a solution of (2) can be found. Thus the following general classes of equations for which we can find a solution of (2) may be noted here.*

1° If $\dfrac{\frac{\partial M}{\partial y}-\frac{\partial N}{\partial x}}{N}$ is a function of x only, say $f_1(x)$, we have, from (2), that $\mu=e^{\int f_1(x)dx}$ is an integrating factor (§ 17).

2° If $\dfrac{\frac{\partial N}{\partial x}-\frac{\partial M}{\partial y}}{M}$ is a function of y only, say $f_2(y)$, then $\mu=e^{\int f_2(y)dy}$ is obviously an integrating factor (§ 17).

3° If the equation is linear, then $M=Py-Q$, $N=1$, and (2) becomes $P\,dx=\dfrac{-P}{Py-Q}\,dy=\dfrac{d\mu}{\mu}$. Hence $\mu=e^{\int P dx}$ is an integrating factor (§ 13).

4° If M and N are homogeneous and of the same degree n, we get, by composition, after we have multiplied numerator and denominator of the first two members of (2) by y and x respectively,

$$(3)\qquad \frac{\left(y\frac{\partial M}{\partial y}-y\frac{\partial N}{\partial x}\right)dx+\left(x\frac{\partial N}{\partial x}-x\frac{\partial M}{\partial y}\right)dy}{xM+yN}=\frac{d\mu}{\mu}.$$

* These were enumerated in Chapter II, in the list of equations for which integrating factors can be found.

By Euler's theorem for homogeneous functions, we have

$$x\frac{\partial M}{\partial x}+y\frac{\partial M}{\partial y}=nM, \quad \therefore y\frac{\partial M}{\partial y}=nM-x\frac{\partial M}{\partial x},$$

$$x\frac{\partial N}{\partial x}+y\frac{\partial N}{\partial y}=nN, \quad \therefore x\frac{\partial N}{\partial x}=nN-y\frac{\partial N}{\partial y},$$

whence (3) may be written

$$\frac{x\left(\frac{\partial M}{\partial x}dx+\frac{\partial M}{\partial y}dy\right)+y\left(\frac{\partial N}{\partial x}dx+\frac{\partial N}{\partial y}dy\right)-n\,(M\,dx+N\,dy)}{x\,M+y\,N}=-\frac{d\mu}{\mu}.$$

But $M\,dx+N\,dy=0$. Hence we may add $(n+1)(M\,dx+N\,dy)$ to the numerator on the left without altering its value. Doing so, we have

$$\frac{d(xM+yN)}{xM+yN}=-\frac{d\mu}{\mu}.$$

Integrating, we have $$\mu=\frac{1}{xM+yN} \text{ (§ 17).}$$

5° If $M=yf_1(xy)$, $N=xf_2(xy)$, then on multiplying numerator and denominator of the first and second members of (2) by y and $-x$ respectively, we have by composition, after obvious reductions,

$$-\frac{d(xM-yN)}{xM-yN}=\frac{d\mu}{\mu}.$$

On integrating, we have $$\mu=\frac{1}{xM-yN} \text{ (§ 17).}$$

81. Non-linear Partial Differential Equations of First Order. Complete, General, Singular Solutions. — We have seen (§ 76) that a primitive

(1) $$\phi(x, y, z, a, b)=0$$

which involves two arbitrary constants gives rise to a unique partial differential equation of the first order

$$(2) \qquad f(x, y, z, p, q)=0.$$

This differential equation is gotten by eliminating a and b from (1) and its derivatives with respect to x and y respectively; viz.,

$$(3) \qquad \begin{cases} \dfrac{\partial \phi}{\partial x}+p\dfrac{\partial \phi}{\partial z}=0, \\ \dfrac{\partial \phi}{\partial y}+q\dfrac{\partial \phi}{\partial z}=0. \end{cases}$$

(1) is spoken of as a solution of (2). Here, of course, a and b are constants. Letting them be parameters, we have, on differentiating (1), and taking account of (3),

$$(4) \qquad \begin{cases} \dfrac{\partial \phi}{\partial a}\dfrac{\partial a}{\partial x}+\dfrac{\partial \phi}{\partial b}\dfrac{\partial b}{\partial x}=0, \\ \dfrac{\partial \phi}{\partial a}\dfrac{\partial a}{\partial y}+\dfrac{\partial \phi}{\partial b}\dfrac{\partial b}{\partial y}=0. \end{cases}$$

These two equations can be consistent only in case either

$$(5) \qquad \begin{cases} \dfrac{\partial \phi}{\partial a}=0, \\ \dfrac{\partial \phi}{\partial b}=0, \end{cases}$$

or

$$\begin{vmatrix} \dfrac{\partial a}{\partial x} & \dfrac{\partial b}{\partial x} \\ \dfrac{\partial a}{\partial y} & \dfrac{\partial b}{\partial y} \end{vmatrix}=0.$$

If this determinant vanishes, b is some function of a, say $\psi(a)$, (Note I in the Appendix). Then (4) may be replaced by

$$(6)\qquad \begin{cases} b=\psi(a), \\ \dfrac{\partial\phi}{\partial a}+\psi'(a)\dfrac{\partial\phi}{\partial b}=0. \end{cases}$$

Since (5) and (6) were gotten on the assumption that (3) hold, it follows that if we eliminate a and b from (1) and (5), or from (1) and (6), we shall get relations between x, y, z which will satisfy (2). Hence these relations are also solutions.

We see then that the primitive of a partial differential equation of the first order is not the only solution. But since the others can be gotten from it, Lagrange called it the *complete* solution.

We have already noted (§ 78) that, in the general theory of partial differential equations, it is proved that an arbitrary function appears in the general solution of an equation of the first order. Since in (6) the function $\psi(a)$ is any function of a we please, Lagrange called the solution gotten by eliminating a and b from (1) and (6) the *general* solution.

A *particular* solution is gotten by assigning definite values to a and b in the complete solution, or by using a definite function $\psi(a)$ and eliminating a and b from (1) and (6).

On the other hand, the solution gotten by eliminating a and b from (1) and (5) contains nothing arbitrary, and is known as the *singular* solution. It is the exact analogue of the singular solution of ordinary differential equations. Looked upon geometrically, it is the equation of the envelope of the doubly infinite number of surfaces whose equation is given by (1), just as the general solution is the envelope of an arbitrarily chosen single infinity of those surfaces. It can also be shown * that the singular solution can be gotten from the differential equation by eliminating p and q from

$$f(x, y, z, p, q)=0,\quad \frac{\partial f}{\partial p}=0,\quad \frac{\partial f}{\partial q}=0,$$

* See Goursat-Bourlet, p. 24, also p. 109 and foll.

in exact analogy to the case of ordinary differential equations of the first order. But the limits of this book prohibit a further discussion of the subject. We shall conclude it with the following obvious remarks:

1° There need be no singular solution. This will happen in case the equations (1) and (5) are inconsistent; or, geometrically, in case the surfaces (1) have no envelope.

2° The general solution cannot be written down. $\psi(a)$ is entirely at our disposal; but until it is given, there is no way of eliminating a and b from (1) and (6), and when it is given, we have, of course, a particular solution.

3° There is no unique complete solution. *Any solution involving two arbitrary constants may be taken as one.* It is easy to see that there is an indefinite number of them. For, if we choose any form for $\psi(a)$ which involves two arbitrary constants h and k, on eliminating a and b from (1) and (6) we get a solution involving these two arbitrary constants, which fulfills all the requirements for a complete solution.

We saw in (§ 76) that a complete solution of $z^2(p^2+q^2+1)=R^2$ is

$$(x-a)^2+(y-b)^2+z^2=R^2,$$

which represents a family of spheres of radius R and centers in the plane $z=0$. The envelope of these is the pair of planes $z^2=R^2$, which is obviously the result of eliminating a and b from

$$(x-a)^2+(y-b)^2+z^2=R^2,\ x-a=0,\ y-b=0,$$

or of eliminating p and q from

$$z^2(p^2+q^2+1)=R^2,\ z^2p=0,\ z^2q=0.$$

$z^2=R^2$ is then a singular solution.

The choice of any relation $b=\psi(a)$ results in selecting those spheres whose centers lie along the curve $y=\psi(x)$ in the plane $z=0$. The envelope of these is obviously the tubular surface traced out by the motion of a sphere of radius R

moving with its center on the curve $y = \psi(x)$ and along its entire length. In particular, if $\psi(x)$ is a linear function of x, the envelope is a cylinder.

As an example, suppose $b = ha + k$.

To find the corresponding solution, we have to eliminate a from

$$(x - a)^2 + (y - ha - k)^2 + z^2 = R^2,$$

$$(x - a) + h(y - ha - k) = 0.$$

From the second of these we have

$$a = \frac{x + hy - hk}{1 + h^2}.$$

Putting this in the other equation, we have

$$\frac{h^2(hx - y + k)^2 + (hx - y + k)^2}{(1 + h^2)^2} = R^2 - z^2,$$

or

$$(hx - y + k)^2 = (1 + h^2)(R^2 - z^2),$$

which is obviously the equation of a circular cylinder whose axis is the line

$$y = hx + k,\ z = 0.$$

Since this solution contains two arbitrary constants, it is a complete solution. (See Ex. 6, § 83.)

As an exercise, the student should start with

$$(hx - y + k)^2 = (1 + h^2)(R^2 - z^2)$$

as a primitive and show that, considering h and k as the arbitrary constants, the resulting partial differential equation is again $z^2(p^2 + q^2 + 1) = R^2$.

82. Method of Lagrange and Charpit. — Since the knowledge of a complete solution is sufficient to enable us to find all other solutions, it is usually desirable, whenever possible, to find it first. Lagrange suggested the following method:

Given the equation

$$(1) \qquad f(x, y, z, p, q) = 0,$$

try to find a second relation

(2) $$\phi(x, y, z, p, q, a) = 0,$$

which involves either p or q or both, and also an arbitrary constant, and such that when we solve (1) and (2) for p and q and put these in the total differential equation

(3) $$dz = p\,dx + q\,dy$$

the latter can be solved; the solution of (3) is a complete solution of (1), since, from what precedes, it determines z as such a function of x and y, that the values of p and q obtained from it, together with z satisfy (1); besides it involves two arbitrary constants. To do this we may proceed as follows:

Differentiating (1) and (2) with respect to x, we have

(4) $$\frac{\partial f}{\partial x} + p\frac{\partial f}{\partial z} + \frac{\partial f}{\partial p}\frac{dp}{dx} + \frac{\partial f}{\partial q}\frac{dq}{dx} = 0,*$$

(5) $$\frac{\partial \phi}{\partial x} + p\frac{\partial \phi}{\partial z} + \frac{\partial \phi}{\partial p}\frac{dp}{dx} + \frac{\partial \phi}{\partial q}\frac{dq}{dx} = 0.$$

Similarly, differentiating with respect to y,

(6) $$\frac{\partial f}{\partial y} + q\frac{\partial f}{\partial z} + \frac{\partial f}{\partial p}\frac{dp}{dy} + \frac{\partial f}{\partial q}\frac{dq}{dy} = 0,$$

(7) $$\frac{\partial \phi}{\partial y} + q\frac{\partial \phi}{\partial z} + \frac{\partial \phi}{\partial p}\frac{dp}{dy} + \frac{\partial \phi}{\partial q}\frac{dq}{dy} = 0.$$

Now the condition that (3) shall be integrable is obviously

(8) $$\frac{dp}{dy} - \frac{dq}{dx} = 0.$$

* Throughout this section, $\frac{d}{dx} \equiv \frac{\partial}{\partial x} + p\frac{\partial}{\partial z}$, $\frac{d}{dy} \equiv \frac{\partial}{\partial y} + q\frac{\partial}{\partial z}$.

From equations (4), (5), (6), (7), (8) we can eliminate the four quantities $\frac{dp}{dx}, \frac{dp}{dy}, \frac{dq}{dx}, \frac{dq}{dy}$. We get rid of $\frac{dp}{dx}$ by multiplying (4) and (5) by $\frac{\partial\phi}{\partial p}$ and $\frac{\partial f}{\partial p}$ respectively, and subtracting. The result is

$$(9)\quad \left(\frac{\partial f}{\partial x}+p\frac{\partial f}{\partial z}\right)\frac{\partial\phi}{\partial p}-\left(\frac{\partial\phi}{\partial x}+p\frac{\partial\phi}{\partial z}\right)\frac{\partial f}{\partial p}+\left(\frac{\partial f}{\partial q}\frac{\partial\phi}{\partial p}-\frac{\partial\phi}{\partial q}\frac{\partial f}{\partial p}\right)\frac{dq}{dx}=0.$$

Multiplying (6) and (7) by $\frac{\partial\phi}{\partial q}$ and $\frac{\partial f}{\partial q}$ respectively, and subtracting, we get rid of $\frac{dq}{dy}$ and have

$$(10)\quad \left(\frac{\partial f}{\partial y}+q\frac{\partial f}{\partial z}\right)\frac{\partial\phi}{\partial q}-\left(\frac{\partial\phi}{\partial y}+q\frac{\partial\phi}{\partial z}\right)\frac{\partial f}{\partial q}+\left(\frac{\partial f}{\partial p}\frac{\partial\phi}{\partial q}-\frac{\partial\phi}{\partial p}\frac{\partial f}{\partial q}\right)\frac{dp}{dy}=0.$$

Making use of (8), we have, on adding and arranging according to derivatives of ϕ,

$$(11)\quad \left(\frac{\partial f}{\partial x}+p\frac{\partial f}{\partial z}\right)\frac{\partial\phi}{\partial p}+\left(\frac{\partial f}{\partial y}+q\frac{\partial f}{\partial z}\right)\frac{\partial\phi}{\partial q}-\frac{\partial f}{\partial p}\frac{\partial\phi}{\partial x}-\frac{\partial f}{\partial q}\frac{\partial\phi}{\partial y}$$

$$-\left(p\frac{\partial f}{\partial p}+q\frac{\partial f}{\partial q}\right)\frac{\partial\phi}{\partial z}=0.$$

This is a linear equation. Hence, to solve for ϕ, we consider the system of ordinary equations

$$(12)\quad \frac{dp}{\frac{\partial f}{\partial x}+p\frac{\partial f}{\partial z}}=\frac{dq}{\frac{\partial f}{\partial y}+q\frac{\partial f}{\partial z}}=\frac{dx}{-\frac{\partial f}{\partial p}}=\frac{dy}{-\frac{\partial f}{\partial q}}=\frac{dz}{-\left(p\frac{\partial f}{\partial p}+q\frac{\partial f}{\partial q}\right)}=\frac{d\phi}{0}.$$

Our aim is to find, not the most general form of ϕ,* but any form of it that contains p or q, and an arbitrary constant. Hence, in actual practice, we look for the simplest one.

Remark. — It is desirable that (11) or (12) be committed to memory.

If we put $$\frac{d}{dx}=\frac{\partial}{\partial x}+p\frac{\partial}{\partial z}, \text{ and } \frac{d}{dy}=\frac{\partial}{\partial y}+q\frac{\partial}{\partial z},$$

we have the skew symmetrical form

(12′) $$\frac{dp}{\frac{df}{dx}}=-\frac{dx}{\frac{\partial f}{\partial p}}=\frac{dq}{\frac{df}{dy}}=-\frac{dy}{\frac{\partial f}{\partial q}};$$

as for the next term $\dfrac{dz}{-\left(p\dfrac{\partial f}{\partial p}+q\dfrac{\partial f}{\partial q}\right)}$, this is really a result of the equality of $\dfrac{dx}{-\dfrac{\partial f}{\partial p}}=\dfrac{dy}{-\dfrac{\partial f}{\partial q}}$, as may be seen by composition after multiplying numerator and denominator of the first by p and of the second by q, and taking account of (3).

Or using the same notation as before, and writing

$$\frac{df}{dx}\frac{\partial\phi}{\partial p}-\frac{\partial f}{\partial p}\frac{d\phi}{dx}=[f,\phi]_{x,p},$$

$$\frac{df}{dy}\frac{\partial\phi}{\partial q}-\frac{\partial f}{\partial q}\frac{d\phi}{dy}=[f,\phi]_{y,q},$$

(11) may be put in the compact form

(11′) $$[f,\phi]_{x,p}+[f,\phi]_{y,q}=0.$$

* Lagrange thought originally that by using the general form of ϕ (which would involve an arbitrary function) he could get the general solution by this method. But this is, in fact, not practicable. Charpit, in a memoir presented to the Académie des Sciences, in June, 1784, suggested the use of any form for ϕ involving p or q and an arbitrary constant, obtaining by its use the complete instead of the general solution.

Ex. 1 $z - pq = 0$.

To find ϕ we must find a solution of

$$\frac{dp}{p} = \frac{dq}{q} = \frac{dx}{q} = \frac{dy}{p} = \cdots.$$

Using the first two members, we have

$$\phi \equiv p - aq = 0.$$

Combining this with the original equation, we have

$$p = a\sqrt{\frac{z}{a}},\ q = \sqrt{\frac{z}{a}}.$$

Then (3) becomes $dz = a\sqrt{\frac{z}{a}}\,dx + \sqrt{\frac{z}{a}}\,dy$,

or $$\frac{\sqrt{a}\,dz}{\sqrt{z}} = a\,dx + dy.$$

Integrating, we have $2\sqrt{az} = ax + y + b$,

or $$4\,az = (ax + y + b)^2,$$

which is a complete solution.

If we had used the second and third members, we would have had

$$\phi \equiv q - x - a = 0.\quad \text{Whence}$$

$$q = x + a, \text{ and } p = \frac{z}{x + a}.$$

Then (3) becomes $dz = \frac{z}{x + a}\,dx + (x + a)\,dy$,

or $$dy = \frac{(x + a)\,dz - z\,dx}{(x + a)^2}.$$

Integrating, we have $y + b = \dfrac{z}{x + a}$,

or $$z = (x + a)(y + b),$$

which is also a complete solution.

The general and singular solutions can be gotten from either of these. Thus, using the second one, we get the general solution by eliminating a from

$$z = (x + a)[y + \psi(a)]$$

and $$y + \psi(a) + \psi'(a)(x + a) = 0,$$

where $\psi(a)$ is any function of a.

In particular let the student show that if we put

$$\psi(a) = k - ha$$

where h and k are any constants, the corresponding solution is $4hz = (hx + y + k)^2$, which we recognize as the first form obtained.

The singular solution, resulting from eliminating a and b from

$$z = (x + a)(y + b),$$

$$y + b = 0,$$

$$x + a = 0,$$

is $z = 0$.

Let the student show that this can also be gotten from the other form of solution.

Ex. 2. $p = (z + yq)^2$.

Ex. 3. $\sqrt{p} + \sqrt{q} = 2x$.

83. Special Methods. — Special methods for certain forms of the differential equation at times prove simpler than the general method of § 82 (although most of them are suggested by the latter). Some of these are the following:

1° Suppose all the variables absent. The equation takes the form

$$f(p, q) = 0, \tag{1}$$

and the equations for determining ϕ become

$$\frac{dp}{0} = \frac{dq}{0} = \cdots. \tag{2}$$

From the first member we see that $p = a$.

Then q is gotten by substituting in (1). Evidently $q = b$, where b is determined by $f(a, b) = 0$.

We have then $$dz = adx + bdy,$$

or $$z = ax + by + c, \text{ where } f(a, b) = 0.$$

Hence the rule:

The complete solution of $f(p, q) = 0$ is

$$z = ax + by + c, \textit{ where } f(a, b) = 0.*$$

By means of simple transformations, certain forms of equations can be brought into this type:

Putting $\log z = Z$, or $z = e^Z$, we have $p = \frac{\partial z}{\partial x} = e^Z \frac{\partial Z}{\partial x} = z \frac{\partial Z}{\partial x}$; so that

$$\frac{p}{z} = \frac{\partial Z}{\partial x}.$$

Similarly $$\frac{q}{z} = \frac{\partial Z}{\partial y}.$$

* The complete solution represents a doubly infinite set of planes. Any particular solution represents the envelope of a chosen single infinity of these. This is a developable surface. There is no singular solution. Why?

Hence, if the equation is $f\left(\frac{p}{z}, \frac{q}{z}\right) = 0$, $\log z = Z$ will transform it into

$$f\left(\frac{\partial Z}{\partial x}, \frac{\partial Z}{\partial y}\right) = 0.$$

Again, if we let $\log x = X$, or $x = e^X$,

$$p = \frac{\partial z}{\partial x} = \frac{\partial z}{\partial X}\frac{dX}{dx} = \frac{1}{x}\frac{\partial z}{\partial X}; \quad i.e.\ xp = \frac{\partial z}{\partial X}.$$

Similarly, putting $\log y = Y$, we get $yq = \frac{\partial z}{\partial Y}$.

Hence

$f(xp, q) = 0$ is transformed into $f\left(\frac{\partial z}{\partial X}, \frac{\partial z}{\partial y}\right) = 0$ by $\log x = X$;

$f(p, yq) = 0$ is transformed into $f\left(\frac{\partial z}{\partial x}, \frac{\partial z}{\partial Y}\right) = 0$ by $\log y = Y$;

$f(xp, yq) = 0$ is transformed into $f\left(\frac{\partial z}{\partial X}, \frac{\partial z}{\partial Y}\right) = 0$ by $\log x = X$, $\log y = Y$;

$f\left(\frac{xp}{z}, \frac{q}{z}\right) = 0$ is transformed into $f\left(\frac{\partial Z}{\partial X}, \frac{\partial Z}{\partial y}\right) = 0$ by $\log x = X$, $\log z = Z$;

and so on.

Ex. 1. $pq = 1$.

The complete solution is $z = ax + by + c$, where $ab = 1$,

or
$$z = ax + \frac{1}{a}y + c.$$

Ex. 2. $q = z + px$.

Writing this in the form $\frac{q}{z} = 1 + \frac{px}{z}$, we see that the substitution $X = \log x$, $Z = \log z$, will transform this equation into

$$\frac{\partial Z}{\partial y} = 1 + \frac{\partial Z}{\partial X}.$$

The complete solution is $Z = aX + (1 + a)y + c$. Passing back to the original variables, we have $\log z = a \log x + (1 + a)y + c$. or

$$z = bx^a e^{(1+a)y}.$$

Ex. 3. $px + qy = 1$.

Ex. 4. $x^2p^2 + y^2q^2 = z^2$.

Ex. 5. $yq = p$.

2° If x and y are absent, the equation takes the form

$$(1) \qquad f(z, p, q) = 0,$$

and we have for determining ϕ

$$(2) \qquad \frac{dp}{p\dfrac{\partial f}{\partial z}} = \frac{dq}{q\dfrac{\partial f}{\partial z}} = \cdots,$$

$$\therefore \frac{dp}{p} = \frac{dq}{q}, \text{ whence } q = ap.$$

Substituting in (1), we have $f(z, p, ap) = 0$, whence $p = \psi(z, a)$, and $dz = p\,dx + q\,dy$ becomes $\dfrac{dz}{\psi(z, a)} = dx + a\,dy$, where the variables are separated.

To put this rule in shape easily to be carried in mind, we note that, to say $q = ap$ is to say that z is a function of $x + ay$, by the general method of § 79. If we put $x + ay = t$, we have

$$p = \frac{\partial z}{\partial x} = \frac{dz}{dt}, \ q = \frac{\partial z}{\partial y} = a\frac{dz}{dt},$$

and the equation (1) becomes the *ordinary* differential equation

$$f\left(z, \frac{dz}{dt}, a\frac{dz}{dt}\right) = 0,$$

in which the independent variable is absent. Hence the variables can be separated immediately after solving for $\frac{dz}{dt}$. We have, then, the rule:

If the equation is of the form $f(z, p, q) = 0$, *put* $x + ay = t$, *which will replace* p *by* $\frac{dz}{dt}$, *and* q *by* $a\frac{dz}{dt}$. *Put these values in the equation and solve for* $\frac{dz}{dt}$.

If the equation has one of the forms

$$f(z, xp, q) = 0, f(z, p, yq) = 0, f(z, xp, yq) = 0,$$

one, or both of the substitutions $\log x = X$, $\log y = Y$ will reduce it to the above form.

Ex. 6. $z^2(p^2 + q^2 + 1) = R^2$.

Putting $x + ay = t$, $p = \frac{dz}{dt}$, $q = a\frac{dz}{dt}$, and the equation becomes

$$z^2\left[\left(\frac{dz}{dt}\right)^2(1 + a^2) + 1\right] = R^2,$$

or

$$\frac{\sqrt{1 + a^2}\,z\,dz}{\sqrt{R^2 - z^2}} = dt.$$

$$\therefore -\sqrt{1 + a^2}\sqrt{R^2 - z^2} = t + b = x + ay + b,$$

or

$$(1 + a^2)(R^2 - z^2) = (x + ay + b)^2.$$

Ex. 7. $xp(1 + q) = qz$.

Ex. 8. $z = pq$.

3° If the dependent variable is absent, and the equation is such that it can be put in the form

(1) $$f_1(x, p) = f_2(y, q)$$

(a sort of separation of the variables), the equations to determine ϕ become

$$(2)\qquad \frac{dp}{\frac{\partial f_1}{\partial x}} = \cdots = \frac{dx}{-\frac{\partial f_1}{\partial p}} = \cdots.$$

$$\therefore \frac{\partial f_1}{\partial x} dx + \frac{\partial f_1}{\partial p} dp = 0, \text{ or } df_1 = 0.$$ Hence we have

$$f_1 = a.$$

$$\therefore f_2 = a.$$

Solving these, we have $p = \psi_1(x, a)$, $q = \psi_2(y, a)$, and $dz = p\,dx + q\,dy$ becomes

$$dz = \psi_1(x, a)dx + \psi_2(y, a)dy,$$

in which the variables are separated. Hence the rule:

If the dependent variable is absent, and the other variables are separable, such that the equation takes the form $f_1(x, p) = f_2(y, q)$, *equate each of these members to a constant, solve the resulting equations for* p *and* q, *and put these values in* $dz = p\,dx + q\,dy$.

If the equation can be put in the form $f_1\left(x, \frac{p}{z}\right) = f_2\left(y, \frac{q}{z}\right)$, the transformation $\log z = Z$ will reduce it to the form above.

Ex. 9. $q = 2yp^2$.

Ex. 10. $2(zx - zy) - p + q = 0$.

4° The equation $z = px + qy + f(p, q)$, which is usually referred to as the *extended Clairaut equation* (§ 27), will obviously be solved if we put $p = a$, $q = b$. We have then the complete solution

$$z = ax + by + f(a, b).$$

While the general method of § 82 applies here, it does not give this simple form of solution. By that method we may use either $p = a$ or $q = b$, but not both simultaneously. As a matter of fact it is an accident if the result of substituting in the differential equation the values of p and q obtained from two solutions of equations (12), § 82 is a complete solution. It does happen, at times, as in the case in question. But there is no certainty that it will, nor is there even a likelihood of it.

Ex. **11.** Solve $z = px + qy + \sqrt{p^2 + q^2 + 1}$, and examine for singular solution.

5° If the equation is of the form $f(x+y, p, q) = 0$, let $q = p + a$. Then $f(x+y, p, p+a) = 0$ gives $p = \phi(x+y, a)$, whence

$$q = \phi(x+y, a) + a,$$

and the equation $dz = p\,dx + q\,dy$ becomes

$$dz = \phi(x+y, a)(dx + dy) + a\,dy.$$

Let the student show that this form of solution is given by the general method of § 82.

If the equation is of the form $f\left(x+y, \frac{p}{z}, \frac{q}{z}\right) = 0$, the transformation $\log z = Z$ will reduce it to the form above.

Ex. **12.** $p(x+y) - q = 0.$

Ex. **13.** $zp(x+y) + p(q-p) = z^2.$

6° If either p or q is absent, the method of solution is obvious by inspection. In the former case integrate considering x as a constant, when the constant of integration will be an arbitrary function of x. In the other case integrate considering y as a constant, when the constant of integration will be an arbitrary function of y. These solutions, involving arbitrary functions, are general solutions.*

* When these equations are of the first degree in the derivative, they are linear equations. The method here given is exactly that of Lagrange for such equations (§ 79).

Ex. **14.** $(x-y)q-(x+z)=0.$

Considering x as a constant, we can write $q=\frac{dz}{dy}$.

$$\therefore \frac{dz}{x+z}-\frac{dy}{x-y}=0.$$

Integrating and taking exponentials of both sides, we have

$$(x+z)(x-y)=\phi(x),$$

where $\phi(x)$ is an arbitrary function of x.

Ex. **15.** $xp^2-2zp+xy=0.$

Ex. **16.** $p+y(z-x)=0.$

Ex. **17.** $y^2(p^2-1)=x^2p^2.$

84. Summary. — Partial differential equations of the first order are divided into two general classes: those which are linear in the derivatives of the dependent variable, and those which are not.

1° For the solution of linear differential equations of the first order the method of Lagrange applies, giving the general solution (§ 79).

2° For the solution of non-linear equations of the first order, the general method of Lagrange and Charpit applies (§ 82), giving a complete solution. From this the other solutions can be gotten (§ 81).

At times the special methods of § 83 are shorter than the general method of § 82.

Sometimes a transformation of variables will help in the solution of an equation.

Ex. **1.** $x^2p + y^2q = z^2$.

Ex. **2.** $q = p^2 + 1$.

Ex. **3.** $x\dfrac{\partial u}{\partial x} + y\dfrac{\partial u}{\partial y} + z\dfrac{\partial u}{\partial z} = xyz$.

Ex. **4.** $z = px + qy + (p+q)^2$.

Ex. **5.** $xypq = z^2$.

Ex. **6.** $y^2zp + x^2zq = y^2x$.

Ex. **7.** $q = xp + p^2$.

Ex. **8.** $(p+q)(px+qy) = 1$.

Ex. **9.** $(x+y)(p+q)^2 + (x-y)(p-q)^2 = 1$. [Let $x + y = u^2$, $x - y = v^2$.]

Ex. **10.** $(p^2+q^2)x - pz = 0$.

Ex. **11.** $(x^2+y^2)(p^2+q^2) = 1$. [Let $x = \rho\cos\theta$, $y = \rho\sin\theta$.]

Ex. **12.** $(y^2+z^2-x^2)p - 2xyq + 2xz = 0$.

Ex. **13.** $q^2 = z^2(p-q)$.

Ex. **14.** $(y-x)(qy-px) = (p-q)^2$. [Let $xy = u$, $x + y = v$.]

Ex. **15.** $z - xp - yq = 2\sqrt{x^2+y^2+z^2}$.

Ex. **16.** $pq = px + qy$.

Ex. **17.** $(y+z+u)\dfrac{\partial u}{\partial x} + (z+u+x)\dfrac{\partial u}{\partial y} + (u+x+y)\dfrac{\partial u}{\partial z}$
$= x + y + z$.

Ex. **18.** Determine a system of surfaces such that the normal at each point makes a constant angle with the plane of xy.

Ex. **19.** Determine a system of surfaces such that the coördinates of the point where the normal meets the plane of xy are proportional to the corresponding coördinates of the point on the surface.

Ex. **20.** Determine a system of surfaces for which the product of the distances of the tangent plane from two fixed points is a constant.

CHAPTER XIV

PARTIAL DIFFERENTIAL EQUATIONS OF HIGHER ORDER THAN THE FIRST

85. Partial Differential Equations of the Second Order, Linear in the Second Derivatives. Monge's Method. — The general type of a partial differential equation of the second order linear in the second derivatives is

$$(1) \qquad Rr + Ss + Tt = V,$$

where R, S, T, V are functions of x, y, z, p, q. Gaspard Monge (1746-1818) suggested a method, which is known by his name, by which a first or intermediary integral is found in the form of a partial differential equation of the first order involving an arbitrary function. The solution of this equation by any of the methods of Chapter XIII, or otherwise, will then give the general solution. While this method applies only in case R, S, T, V satisfy certain conditions, it works sufficiently frequently to justify our giving here at least the rule by which solutions are gotten by this method.* Besides

$$(2) \qquad dz = p\,dx + q\,dy,$$

we have

$$(3) \qquad \begin{cases} dp = r\,dx + s\,dy, \\ dq = s\,dx + t\,dy. \end{cases}$$

Eliminating r and t from (1) and (3) we have

$$(4) \quad s(R\,dy^2 - S\,dx\,dy + T\,dx^2) - (R\,dy\,dp + T\,dx\,dq - V\,dx\,dy) = 0.$$

* For a detailed account of this subject see Forsyth, p. 358 and foll., or Boole, Chapter XV.

Whenever it is possible to satisfy simultaneously,

$$(5)\qquad R\,dy^2 - S\,dx\,dy + T\,dx^2 = 0,$$

$$(6)\qquad R\,dy\,dp + T\,dx\,dq - V\,dx\,dy = 0,\text{*}$$

(4) will be satisfied and, therefore, so will (1). (5) is equivalent to two equations of the first order,

$$(7)\quad dy - W_1(x, y, z, p, q)\,dx = 0,\ dy - W_2(x, y, z, p, q)\,dx = 0,$$

which become identical in case

$$(8)\qquad 4\,RT = S^2.$$

Equations (2) and (6), together with either one of (7), constitute a system of three total differential equations in the five variables x, y, z, p, q. Such a system can be solved only in case certain conditions are fulfilled, and it is for this reason that Monge's method does not always work. It will work if we can find two independent solutions of this system

$$u_1(x, y, z, p, q) = c_1,\quad u_2(x, y, z, p, q) = c_2.$$

In this case it turns out that

$$(9)\qquad u_1 = \phi(u_2),$$

where ϕ is an arbitrary function, is an *intermediary* (or *intermediate*) *integral.* Looked upon as a partial differential equation of the first order, (9) must be integrated again. Its general solution will be the solution of (1).

In case it happens that not only one of (7), but each one, together with (2) and (6), determines a system that can be solved, we have

* Equations (5) and (6) are usually referred to as *Monge's equations.*

two intermediary integrals (9). Solving these for p and q, we put the values of the latter in $dz = p\,dx + q\,dy$. The integral of this is the general solution of (1).

Ex. 1. $q^2r - 2pqs + p^2t = 0.$

Monge's equations are

$$q^2\,dy^2 + 2pq\,dx\,dy + p^2\,dx^2 = 0,$$

$$q^2\,dy\,dp + p^2\,dx\,dq = 0.$$

The first of these is a perfect square,

$$(q\,dy + p\,dx)^2 = 0.$$

Substituting this in the second one, it becomes

$$q\,dp - p\,dq = 0,$$

whence $$\frac{p}{q} = c_1.$$

The first one, combined with $dz = p\,dx + q\,dy$, gives

$$dz = 0,$$

whence $$z = c_2.$$

Hence an intermediary integral is

$$p = q\phi(z).$$

In this case we have only one intermediary integral, hence we must integrate this. Since it is linear, it can be solved by the method of Lagrange (§ 79). Its general solution is

$$x\phi(z) + y = \psi(z).$$

Ex. 2. $r - a^2t = 0$.*

Monge's equations are

$$dy^2 - a^2\,dx^2 = 0, \text{ or } dy - a\,dx = 0 \text{ and } dy + a\,dx = 0,$$

$$dy\,dp - a^2\,dx\,dq = 0.$$

Using $dy - a\,dx = 0$, we have $y - ax = c_1$.

Combining this with the second of Monge's equations, we get

$$dp - a\,dq = 0\,; \text{ whence } p - aq = c_2.$$

Hence an intermediary integral is

$$p - aq = \psi(y - ax).$$

Using the other equation, $dy + a\,dx = 0$, we get a second intermediary integral

$$p + aq = \phi(y + ax).$$

Solving these for p and q, we have

$$p = \frac{1}{2}[\phi(y + ax) + \psi(y - ax)],$$

$$q = \frac{1}{2a}[\phi(y + ax) - \psi(y - ax)].$$

* A much simpler method of solution for this equation will be given in § 88. This equation plays an important rôle in Mathematical Physics. It was first integrated by Jean-le-Rond D'Alembert (1717–1783) in a memoir entitled *Récherches sur les vibrations des cordes sonores*, presented in 1747 to the Berlin Academy. In studying the vibrations of a stretched elastic string, he considered the equation in the form $\frac{\partial^2 y}{\partial t^2} - \frac{\partial^2 y}{\partial x^2} = 0$, where t is the time, and x and y are the rectangular coördinates of a point of the string, x the coördinate measured along the line joining the extremities of the string, and y the displacement of the point from the position of equilibrium. His proof is given in Marie, *Histoire des Sciences Mathématiques et Physiques*, t. VIII. p. 217.

We have now to solve

$$dz = \frac{1}{2}[\phi(y + ax) + \psi(y - ax)]dx + \frac{1}{2a}[\phi(y + ax) - \psi(y - ax)]dy$$

$$= \frac{1}{2}\phi(y + ax)(dy + a\,dx) + \frac{1}{2}\psi(y - ax)(dy - a\,dx), \text{ which is exact.}$$

Since ϕ and ψ are symbols of arbitrary functions, we shall retain them in writing the solution

$$z = \phi(y + ax) + \psi(y - ax).$$

There is no loss in failing to add an arbitrary constant, since either of the arbitrary functions may be supposed to incorporate that.

Ex. 3. $r - t = -\dfrac{4p}{x + y}$.

Monge's equations are

$$dy^2 - dx^2 = 0, \text{ or } dy - dx = 0 \text{ and } dy + dx = 0,$$

$$dy\,dp - dx\,dq + \frac{4p}{x + y}dx\,dy = 0.$$

Using $dy - dx = 0$, we have $y - x = c_1$.

Combining this with the second of Monge's equations, we have

$$2x\,dp + 4p\,dx - 2x\,dq + c_1(dp - dq) = 0.$$

Also $dz = p\,dx + q\,dy$ becomes

$$dz = p\,dx + q\,dx.$$

Subtracting twice this from the above equation, we get

$$2(x\,dp + p\,dx) - 2(x\,dq + q\,dx) + c_1(dp - dq) + 2\,dz = 0.$$

This is exact, and has for solution

$$(2x+c_1)(p-q)+2z=c_2,$$

or

$$(x+y)(p-q)+2z=c_2.$$

Hence an intermediary integral is

$$(x+y)(p-q)+2z=\phi(y-x).$$

Using the equation $dy+dx=0$, we get a system of total differential equations which are not integrable. Hence we must integrate the intermediary integral. This is linear, so Lagrange's method applies,

$$\frac{dx}{x+y}=\frac{-dy}{x+y}=\frac{dz}{\phi(y-x)-2z}.$$

From the equation of the first two members we have

$$x+y=a.$$

Replacing y by its value $a-x$, we have

$$\frac{dx}{a}=\frac{dz}{\phi(a-2x)-2z},$$

or

$$\frac{dz}{dx}+\frac{2}{a}z=\frac{1}{a}\phi(a-2x).$$

This is a linear ordinary equation of the first order. An integrating factor (§ 13) is $e^{\frac{2}{a}\int dx}=e^{\frac{2x}{a}}$, and the solution is

$$aze^{\frac{2x}{a}}=\int e^{\frac{2x}{a}}\phi(a-2x)\,dx+b.$$

Replacing a by its value $x+y$, we have the general solution

$$(x+y)ze^{\frac{2x}{x+y}}-\int e^{\frac{2x}{a}}\phi(a-2x)dx=\psi(x+y).$$

Here, as in the case of the non-linear partial differential equations of the first order (§ 81), the general solution cannot be written down. For until the form of ϕ is known the above integral cannot be calculated. In any example in which the initial conditions determine ϕ, the a which appears in the integral must be replaced by $x+y$, after the integration has been effected.

Ex. 4. $q(1+q)r-(p+q+2pq)s+p(1+p)t=0.$

Ex. 5. $ps-qr=0.$

Ex. 6. $(b+cq)^2r-2(b+cq)(a+cp)s+(a+cp)^2t=0.$

86. Special Method. — At times, by considering one or the other of the independent variables as a constant temporarily, the equation may be looked upon as an ordinary differential equation. Of course, an arbitrary function of the variable supposed constant must take the place of the arbitrary constant in the solution. The following examples will illustrate:

Ex. 1. $xr=p.$

Letting y be a constant temporarily, this may be written

$$x\frac{dp}{dx}=p, \text{ or } \frac{dp}{p}=\frac{dx}{x}.$$

Integrating, we have $p=xf(y)$, where $f(y)$ is an arbitrary function. Again letting y be constant, we have

$$\frac{dz}{dx}=xf(y),$$

whence $z=x^2f(y)+\phi(y)$, where $\phi(y)$ is another arbitrary function. Here the factor $\frac{1}{2}$, arising on the right, is incorporated in $f(y)$.

Ex. 2. $r+s+p=0$.

Integrating this, considering y as a constant, we have

$$p+q+z=f(y).$$

This is linear and of the first order. Hence Lagrange's method applies.

$$\frac{dx}{1}=\frac{dy}{1}=\frac{dz}{f(y)-z}.$$

From the first two members we get

$$x-y=a.$$

From the last two members we have the linear ordinary differential equation

$$\frac{dz}{dy}+z=f(y).$$

An integrating factor is e^y (§ 13), and the solution is

$$ze^y=\int f(y)e^y dy+b$$

$$=\phi(y)+b.$$

Hence the general solution is

$$ze^y-\phi(y)=\psi(x-y),$$

or $$z=\phi(y)+e^{-y}\psi(x-y),$$

where the factor e^{-y} is incorporated in $\phi(y)$.

Ex. 3. $yt-q=xy^2$.

Ex. 4. $s=xy$.

Ex. 5. $r+p=xy$.

87. General Linear Partial Differential Equations.—We shall consider now partial differential equations which are linear in the dependent variable and all of its derivatives. The general type of such equations is

$$(1)\quad P_{n,0}\frac{\partial^n z}{\partial x^n}+P_{n-1,1}\frac{\partial^n z}{\partial x^{n-1}\,\partial y}+P_{n-2,2}\frac{\partial^n z}{\partial x^{n-2}\,\partial y^2}+\cdots+P_{0,n}\frac{\partial^n z}{\partial y^n}$$

$$+P_{n-1,0}\frac{\partial^{n-1} z}{\partial x^{n-1}}+\cdots+P_{s,r}\frac{\partial^{r+s} z}{\partial x^s\,\partial y^r}+\cdots+P_{1,0}\frac{\partial z}{\partial x}$$

$$+P_{0,1}\frac{\partial z}{\partial y}+P_{0,0}z=f(x,\ y),$$

where the coefficients are functions of x and y, including the case where some or all of them are constants.

If we put $D=\frac{\partial}{\partial x}$, $\mathcal{D}=\frac{\partial}{\partial y}$, (1) may be written

$$(P_{n,0}D^n+P_{n-1,1}D^{n-1}\mathcal{D}+P_{n-2,2}D^{n-2}\mathcal{D}^2+\cdots+P_{0,n}\mathcal{D}^n+P_{n-1,0}D^{n-1}$$

$$+\cdots+P_{s,r}D^s\mathcal{D}^r+\cdots+P_{1,0}D+P_{0,1}\mathcal{D}+P_{0,0})z=f(x,y),$$

or more briefly

$$(1)\qquad F(D,\mathcal{D})z=f(x,y),$$

where $F(D,\mathcal{D})$ is a symbolic operator, which, looked upon algebraically, is a polynomial of degree n in D and $\mathcal{D}$. There are many points of similarity between this equation and the linear ordinary differential equation of the nth order (§ 42).

Obviously, $F(D,\mathcal{D})(u+v)=F(D,\mathcal{D})u+F(D,\mathcal{D})v.$

Hence the problem of solving (1) can be divided into two, viz. that of finding the general integral of

$$(2) \qquad F(D, \mathcal{D})\, z = 0,$$

which we shall call the *complementary function* of (1), and that of finding any particular integral. The sum of these will give the general integral of (1).

88. Homogeneous Linear Equations with Constant Coefficients. — Following a generally adopted convention, we shall use the term *homogeneous* to apply to an equation in which all the derivatives are of the same order. In this case the symbolic operator is homogeneous in D and $\mathcal{D}$. Suppose, besides, that the coefficients are constants, and the right-hand member zero. Our equation will be of the form

$$(1) \qquad (k_0 D^n + k_1 D^{n-1}\mathcal{D} + \cdots + k_{n-1} D\mathcal{D}^{n-1} + k_n \mathcal{D}^n)\, z = 0,$$

or

$$F(D, \mathcal{D})\, z = 0.$$

Since for ϕ any function whatever

$$D^r \mathcal{D}^s \phi(y + mx) = m^r \phi^{(r+s)}(y + mx),$$

where $\phi^{(r+s)}(y + mx)$ means $\dfrac{d^{r+s}\phi(y + mx)}{[d(y + mx)]^{r+s}}$, the result of substituting $z = \phi(y + mx)$ in (1) will be

$$\phi^{(n)}(y + mx) F(m, 1) = 0.$$

Hence $z = \phi(y + mx)$ will be a solution, provided $F(m, 1) = 0$; *i.e.*

$$(2) \qquad k_0 m^n + k_1 m^{n-1} + \cdots + k_{n-1} m + k_n = 0.$$

If the roots of (2), which we shall speak of as the *auxiliary equation*, are distinct, say $m_1, m_2, \cdots, m_n$,

$$z = \phi_1(y + m_1x) + \phi_2(y + m_2x) + \cdots + \phi_n(y + m_nx)$$

will be a solution. Since it contains n arbitrary functions, it will be the general solution.*

Thus let us consider the equation in Ex. 2, § 85,

$$\frac{\partial^2 z}{\partial x^2} - a^2 \frac{\partial^2 z}{\partial y^2} = 0.$$

The auxiliary equation in this case is

$$m^2 - a^2 = 0. \quad \therefore m = \pm a.$$

Hence the general solution is

$$z = \phi(y + ax) + \psi(y - ax).$$

Ex. 1. $\frac{\partial^2 z}{\partial x^2} - 3 \frac{\partial^2 z}{\partial x \, \partial y} + 2 \frac{\partial^2 z}{\partial y^2} = 0.$

Ex. 2. $\frac{\partial^3 z}{\partial x^3} - 7 \frac{\partial^3 z}{\partial x^2 \, \partial y} + 10 \frac{\partial^3 z}{\partial x \, \partial y^2} = 0.$

89. Roots of Auxiliary Equation Repeated. — If any of the roots of the auxiliary equation are repeated, the method of § 88 fails to give us the general solution. In this case we proceed by a method entirely analogous to that in § 47.

* If $F(D, \mathcal{D})$ contains $\mathcal{D}$ as a factor, $F(m, 1)$ is only of degree $n-1$. The lost root in this case is ∞, and the corresponding integral is $\phi(x)$. This is also obvious from the form of the differential equation in this case. For to say that $\mathcal{D}$ is a factor of $F(D, \mathcal{D})$ is to say that every derivative of z is taken at least once with respect to y. Hence $z = \phi(x)$ will give 0. Similarly, if $F(D, \mathcal{D})$ contains $\mathcal{D}^r$ as a factor, $\phi_1(x)$, $y\phi_2(x)$, $\cdots$, $y^{r-1}\phi_r(x)$ are readily seen to be integrals.

The symbolic operator $F(D, \mathcal{D})$ may be written as the product of its factors

$$(D - m_1\mathcal{D})(D - m_2\mathcal{D}) \cdots (D - m_n\mathcal{D}).$$

Moreover, it is readily seen (§ 46) that the order of these factors is immaterial. Suppose m_1 a repeated root. We wish to find a solution of

$$(D - m_1\mathcal{D})(D - m_1\mathcal{D})z = 0.$$

Putting $(D - m_1\mathcal{D})z = v$, our equation is

$$(D - m_1\mathcal{D})v = 0.$$

Hence, by the method of § 88 $\qquad v = \phi(y + m_1 x).$

We now have to solve $\qquad (D - m_1\mathcal{D})z = \phi(y + m_1 x).$

This is linear and of the first order,

$$p - m_1 q = \phi(y + m_1 x).$$

Hence the method of Lagrange (§ 79) applies,

$$\frac{dx}{1} = -\frac{dy}{m_1} = \frac{dz}{\phi(y + m_1 x)}.$$

From the first two members we have

$$y + m_1 x = a.$$

Putting this in the last member, we have

$$dx = \frac{dz}{\phi(a)}.$$

$$\therefore x\phi(a) - z = b.$$

Hence the general solution is

$$x\phi(y + m_1 x) - z = \psi(y + m_1 x),$$

or $\qquad z = \psi(y + m_1 x) + x\phi(y + m_1 x).$

In other words, if m_1 is a double root of the auxiliary equation, not only is $\phi(y+m_1x)$ an integral, but so also is $x\psi(y+m_1x)$. In an entirely analogous manner it can be shown that if m_1 is an r-fold root,

$$\phi_1(y+m_1x),\ x\phi_2(y+m_1x),\ x^2\phi_3(y+m_1x),\ \cdots,\ x^{r-1}\phi_r(y+m_1x)$$

are all integrals.

Ex. 1. $\dfrac{\partial^2 z}{\partial x^2}-2\dfrac{\partial^2 z}{\partial x\,\partial y}+\dfrac{\partial^2 z}{\partial y^2}=0.$

Ex. 2. $\dfrac{\partial^3 z}{\partial x^3}+\dfrac{\partial^3 z}{\partial x^2\partial y}-\dfrac{\partial^3 z}{\partial x\partial y^2}-\dfrac{\partial^3 z}{\partial y^3}=0.$

Ex. 3. $\dfrac{\partial^4 z}{\partial x^4}-2\dfrac{\partial^4 z}{\partial x^2\partial y^2}+\dfrac{\partial^4 z}{\partial y^4}=0.$

90. Roots of Auxiliary Equation Complex. — If the coefficients in the differential equation are real, the complex roots of the auxiliary equation occur in pairs of conjugates. Hence if $\alpha+i\beta$ is a root, $\alpha-i\beta$ will also be one. The corresponding terms in the complementary function will be

$$\phi(y+\alpha x+i\beta x)+\psi(y+\alpha x-i\beta x).$$

ϕ_1 and ψ_1 being any two arbitrarily chosen functions, there is no loss in putting

$$\phi=\phi_1+i\psi_1,\quad \psi=\phi_1-i\psi_1.$$

Our expression above becomes then

$$\phi_1(y+\alpha x+i\beta x)+\phi_1(y+\alpha x-i\beta x)$$
$$+\,i[\psi_1(y+\alpha x+i\beta x)-\psi_1(y+\alpha x-i\beta x)].$$

For ϕ_1 and ψ_1 any real functions, this is real.

Ex. $\frac{\partial^2 z}{\partial x^2} - 2\frac{\partial^2 z}{\partial x\,\partial y} + 2\frac{\partial^2 z}{\partial y^2} = 0.$

The auxiliary equation is

$$m^2 - 2m + 2 = 0. \quad \therefore m = 1 \pm i.$$

The general solution is

$$z = \phi(y + x + ix) + \psi(y + x - ix).$$

It will assume a real form

$$z = \phi_1(y + x + ix) + \phi_1(y + x - ix)$$
$$+ i[\psi_1(y + x + ix) - \psi_1(y + x - ix)],$$

for ϕ_1 and ψ_1 any real functions.

For example, if, in particular, we choose $\phi_1(u)$ to be $\cos u$, and $\psi_1(u)$ to be e^u, we have

$$\cos(y + x + ix) = \cos(x + y)\cos ix - \sin(x + y)\sin ix$$
$$= \cos(x + y)\cosh x - i\sin(x + y)\sinh x,$$
$$\cos(y + x - ix) = \cos(x + y)\cos ix + \sin(x + y)\sin ix$$
$$= \cos(x + y)\cosh x + i\sin(x + y)\sinh x.$$
$$e^{y+x+ix} - e^{y+x-ix} = e^{y+x}(e^{ix} - e^{-ix}) = 2ie^{y+x}\sin x.$$
$$\therefore z = 2\cos(x + y)\cosh x - 2e^{x+y}\sin x.$$

91. Particular Integral. — General methods for finding the particular integral which must be added to the complementary function to get the general integral, in case the right-hand member of the equation is different from zero, may be deduced along lines entirely analogous to those for linear ordinary differential equations with constant coefficients (§§ 47, 48).* In a large number of cases, these can be found more simply by trial, by methods similar to that of undetermined coefficients (§ 50). The following examples will illustrate:

Ex. 1. $$\frac{\partial^2 z}{\partial x^2}+\frac{\partial^2 z}{\partial x\,\partial y}-2\frac{\partial^2 z}{\partial y^2}=\sin(x+2y)-2\sin(x+y)+x+xy.$$

The complementary function is $\phi(y+x)+\psi(y-2x)$.

To get $\sin(x+2y)$, since all the derivatives are of the second order, we try $z=a\sin(x+2y)$.

Then $\dfrac{\partial^2 z}{\partial x^2}+\dfrac{\partial^2 z}{\partial x\,\partial y}-2\dfrac{\partial^2 z}{\partial y^2}=5a\sin(x+2y)$.

If $a=\frac{1}{5}$, this becomes $\sin(x+2y)$. Hence the required particular integral is $\frac{1}{5}\sin(x+2y)$.

Since $\sin(x+y)$ is part of the complementary function, there is no use in trying $z=b\sin(x+y)$. Trying $z=bx\sin(x+y)$ we get $3b\cos(x+y)$. Hence we must try $z=bx\cos(x+y)$. Doing this, we have as a result of substituting in the equation $-3b\sin(x+y)$. This will be $-2\sin(x+y)$ if $b=\frac{2}{3}$. Hence $\frac{2}{3}x\cos(x+y)$ is the required particular integral. [It is obvious that we might have also used $z=by\cos(x+y)$.] To get x, we try $z=cx^3$. Substituting this, we get $6cx$. This equals x if $6c=1$; hence the corresponding particular integral is $\dfrac{x^3}{6}$. To get xy, we try $z=fx^3y$. Using this we get $6fxy+3fx^2$. So we try $z=fx^3y+gx^4$. Using this, we get $6fxy+(3f+12g)x^2$. This equals xy if $f=\frac{1}{6}$, $g=-\frac{1}{24}$. Hence the required particular integral is $\frac{1}{6}x^3y-\frac{1}{24}x^4$. And the general solution is

$$z=\phi(y+x)+\psi(y-2x)+\tfrac{1}{5}\sin(x+2y)+\tfrac{2}{3}x\cos(x+y)+\tfrac{1}{6}x^3+\tfrac{1}{6}x^3y-\tfrac{1}{24}x^4.$$

* Thus, for example, see Forsyth, § 250, Johnson, § 320 and foll., Murray, § 130.

Ex. 2. $\frac{\partial^2 z}{\partial x^2} - 3\frac{\partial^2 z}{\partial x\,\partial y} + 2\frac{\partial^2 z}{\partial y^2} = e^{x+2y} + e^{x+y}.$

The complementary function is $\phi(y+x) + \psi(y+2x)$.

To get e^{x+2y} we try $z = ae^{x+2y}$. Using this, we get $3\,ae^{x+2y}$.

Hence $\frac{1}{3}e^{x+2y}$ is the particular integral desired.

Since e^{x+y} is a part of the complementary function, let us try $z = bxe^{x+y}$. Using this, we get $-be^{x+y}$. Hence $-xe^{x+y}$ is the particular integral desired. [Of course, we might have used $z = bye^{x+y}$ instead.] And the general solution is

$$z = \phi(y+x) + \psi(y+2x) + \tfrac{1}{3}e^{x+2y} - xe^{x+y}.$$

Ex. 3. $\frac{\partial^3 z}{\partial x^2\,\partial y} - 2\frac{\partial^3 z}{\partial x\,\partial y^2} + \frac{\partial^3 z}{\partial y^3} = \frac{1}{x^2}.$

The auxiliary equation is $m^2 - 2m + 1 = 0$. $\therefore m = 1, 1$.

[We have here an example of the case cited in the footnote of § 88.]

The complementary function is $\phi(x) + \psi(y+x) + x\chi(y+x)$.

Since there is no term in $\frac{\partial^3 z}{\partial x^3}$, in order to get $\frac{1}{x^2}$, we have to take y times a function of x which on being differentiated twice gives $\frac{1}{x^2}$; that is, we shall try $z = ay\log x$. Doing this, we find that if $a = -1$ we get $\frac{1}{x^2}$.

Hence the general solution is

$$z = \phi(x) + \psi(y+x) + x\chi(y+x) - y\log x.$$

Ex. 4. $2\frac{\partial^2 z}{\partial x^2} - \frac{\partial^2 z}{\partial x\,\partial y} - 3\frac{\partial^2 z}{\partial y^2} = \sin(x+y) + 3x^2y.$

Ex. 5. $\frac{\partial^2 z}{\partial x^2} - \frac{\partial^2 z}{\partial x\,\partial y} - 2\frac{\partial^2 z}{\partial y^2} = x - y.$

92. Non-Homogeneous Linear Equations with Constant Coefficients. — If $F(D, \mathcal{D})$ of § 87 is not homogeneous in D and $\mathcal{D}$, but the coefficients are constants, it is only under certain circumstances that we can obtain solutions involving arbitrary functions, although we can always find solutions with an indefinite number of arbitrary constants.

Since $D^r e^{ax+by} = a^r e^{ax+by}$, $\mathcal{D}^s e^{ax+by} = b^s e^{ax+by}$, the result of substituting $z = ce^{ax+by}$ in the left-hand member of

$$(1) \qquad F(D, \mathcal{D})z = 0$$

is $ce^{ax+by} F(a, b)$. If a and b satisfy the relation

$$(2) \qquad F(a, b) = 0,$$

which we shall call the *auxiliary equation*, $z = ce^{ax+by}$ will be a solution of (1) where c is any constant. Corresponding to any value of b there will be a definite number of values of a satisfying (2). Hence we can find as many particular solutions as we please by giving various values to b. Now, the sum of any number of integrals of (1) is also an integral. Hence

$$(3) \qquad z = \Sigma\, ce^{ax+by}$$

is a solution where the c's and b's are arbitrary constants, indefinite in number, and each a is so chosen that with the corresponding b it satisfies (2).

If corresponding to any value of b we have the k values of a satisfying (2) in the form $f_1(b), f_2(b), \cdots, f_k(b)$, we can write (3) in the form

$$(4) \qquad z = \Sigma ce^{f_1(b)x+by} + \Sigma ce^{f_2(b)x+by} + \cdots + \Sigma ce^{f_k(b)x+by},$$

where the c's and the b's are perfectly arbitrary.

In general, $F(D, \mathcal{D})$ has no rational factors. If there is a linear factor $D - \lambda\mathcal{D} - \mu$, the equation (2) will contain the factor

$a - \lambda b - \mu$, whence $a = \lambda b + \mu$. Hence one of the f's, say $f_1(b)$, becomes $\lambda b + \mu$, and the corresponding set of terms in (4) may be written

(5) $$z = \Sigma c e^{b(\lambda x + y) + \mu x} = e^{\mu x} \Sigma c e^{b(\lambda x + y)}.$$

Since the c's and the b's are arbitrary, $\Sigma c e^{b(\lambda x + y)}$ is an arbitrary function of $\lambda x + y$, say $\phi(\lambda x + y)$. So that (5) may be written

(6) $$z = e^{\mu x} \phi(\lambda x + y).$$

Hence we see that corresponding to every distinct linear factor of (2) we have a solution of the form (6).*

If there is a linear factor of $F(D, \mathcal{D})$ which is free of D, the corresponding factor in (2) will be free of a; let it be $b - \mu$. The corresponding set of terms in (4) may be written

(5′) $$z = \Sigma c e^{ax + \mu y} = e^{\mu y} \Sigma c e^{ax}.$$

Since the c's and a's are arbitrary, $\Sigma c e^{ax}$ is an arbitrary function of x, and (5′) may be written

(6′) $$z = e^{\mu y} \phi(x).$$

If the right-hand member of the differential equation is not zero, a particular integral may frequently be gotten by trial as in § 91.

Ex. 1. $$\frac{\partial^2 z}{\partial x^2} + \frac{\partial^2 z}{\partial x \, \partial y} + \frac{\partial z}{\partial y} - z = \sin(x + 2y).$$

The auxiliary equation (2) is

$$a^2 + ab + b - 1 = 0,$$

or $$(a + 1)(a + b - 1) = 0.$$

* In the case of the homogeneous equations (§ 88) all the factors are linear. Besides, in that case, $\mu = 0$, and the result there obtained coincides with what we have found here.

Using $a+1=0$, we have $\lambda=0$, $\mu=-1$, and from (6) we see that $z=e^{-x}\phi(y)$ is a solution.

Using $a+b-1=0$, we have $\lambda=-1, \mu=1$. Hence $z=e^{x}\psi(y-x)$ is a solution. The complementary function is, therefore,

$$e^{-x}\phi(y)+e^{x}\psi(y-x).$$

For a particular integral try $z=\alpha\sin(x+2y)+\beta\cos(x+2y)$.

Substituting this in the left-hand member, we get

$$(-4\alpha-2\beta)\sin(x+2y)+(-4\beta+2\alpha)\cos(x+2y).$$

This will equal $\sin(x+2y)$ if $\alpha=-\frac{1}{5}$, $\beta=-\frac{1}{10}$. Hence a particular integral is

$$-\frac{1}{5}\sin(x+2y)-\frac{1}{10}\cos(x+2y),$$

and the general solution is

$$z=e^{-x}\phi(y)+e^{x}\psi(y-x)-\frac{1}{10}[2\sin(x+2y)+\cos(x+2y)].$$

Ex. 2. $\dfrac{\partial^2 z}{\partial x^2}-\dfrac{\partial^2 z}{\partial y^2}+2\dfrac{\partial z}{\partial x}+z=e^{-x}$.

The auxiliary equation (2) is

$$a^2-b^2+2a+1=0$$

or $$(a+b+1)(a-b+1)=0.$$

Hence the complementary function is

$$e^{-x}[\phi(y-x)+\psi(y+x)].$$

Since e^{-x} is part of the complementary function, we would naturally try $z = \alpha x e^{-x}$. But this is also part of the complementary function, as may be seen by putting $\phi = -\frac{y-x}{2}$, $\psi = \frac{y+x}{2}$.

We must then try $z = \alpha x^2 e^{-x}$. Substituting this in the left-hand member, we get $2\alpha e^{-x}$. Hence $\alpha = \frac{1}{2}$, and the particular integral desired is $\frac{1}{2}x^2e^{-x}$. The general solution is

$$z = e^{-x}\left[\phi(y-x) + \psi(y+x) + \frac{1}{2}x^2\right].$$

Ex. 3. $\frac{\partial^2 z}{\partial x^2} - \frac{\partial^2 z}{\partial x\,\partial y} - 2\frac{\partial^2 z}{\partial y^2} + 2\frac{\partial z}{\partial x} + 2\frac{\partial z}{\partial y} = e^{2x+3y} + \sin(2x+y).$

Ex. 4. $\frac{\partial^2 z}{\partial x^2} - \frac{\partial^2 z}{\partial x\,\partial y} + \frac{\partial z}{\partial y} - z = \cos(x+2y) + e^{y}.$

93. Equations Reducible to Linear Equations with Constant Coefficients. — If the coefficient of $D^r \mathcal{D}^s$ in $F(D, \mathcal{D})$ of § 87 is a constant times $x^r y^s$, the equation can be reduced to one with constant coefficients by the transformation $\log x = X$, $\log y = Y$. (Compare with Cauchy's equation, § 51.) Thus

$$Dz = \frac{\partial z}{\partial x} = \frac{1}{x}\frac{\partial z}{\partial X}, \qquad \therefore\ xDz = \frac{\partial z}{\partial X}.$$

$$D^2 z = \frac{\partial^2 z}{\partial x^2} = \frac{1}{x^2}\frac{\partial^2 z}{\partial X^2} - \frac{1}{x^2}\frac{\partial z}{\partial X}, \qquad \therefore\ x^2D^2z = \frac{\partial^2 z}{\partial X^2} - \frac{\partial z}{\partial X}.$$

$$\mathcal{D}z = \frac{\partial z}{\partial y} = \frac{1}{y}\frac{\partial z}{\partial Y}, \qquad \therefore\ y\mathcal{D}z = \frac{\partial z}{\partial Y}.$$

$$\mathcal{D}^2 z = \frac{\partial^2 z}{\partial y^2} = \frac{1}{y^2}\frac{\partial^2 z}{\partial Y^2} - \frac{1}{y^2}\frac{\partial z}{\partial Y}, \qquad \therefore\ y^2\mathcal{D}^2z = \frac{\partial^2 z}{\partial Y^2} - \frac{\partial z}{\partial Y}.$$

$$D\mathcal{D}z = \frac{\partial^2 z}{\partial x\,\partial y} = \frac{1}{xy}\frac{\partial^2 z}{\partial X\,\partial Y}, \qquad \therefore\ xyD\mathcal{D}z = \frac{\partial^2 z}{\partial X\,\partial Y}.$$

. . . .

Ex. 1. $x^2\frac{\partial^2 z}{\partial x^2}+2xy\frac{\partial^2 z}{\partial x\,\partial y}+y^2\frac{\partial^2 z}{\partial y^2}=x^2+y^2.$

Making the substitution $\log x = X$, $\log y = Y$, the equation becomes $\frac{\partial^2 z}{\partial X^2}+2\frac{\partial^2 z}{\partial X\,\partial Y}+\frac{\partial^2 z}{\partial Y^2}-\frac{\partial z}{\partial X}-\frac{\partial z}{\partial Y}=e^{2X}+e^{2Y}$, which has constant coefficients. The auxiliary equation, (2), § 92, is

$$a^2+2ab+b^2-a-b=0, \text{ or } (a+b)(a+b-1)=0.$$

Hence the complementary function is $\phi(Y-X)+e^X\psi(Y-X)$.

For the particular integral try $z=\alpha e^{2X}+\beta e^{2Y}$. Doing this, we find that $\alpha=\beta=\frac{1}{2}$. Hence the general solution is

$$z=\phi(Y-X)+e^X\psi(Y-X)+\frac{1}{2}(e^{2X}+e^{2Y}).$$

Passing back now to x and y, and remembering that $Y-X=\log\frac{y}{x}$, the general solution takes the form

$$z=\phi\left(\frac{y}{x}\right)+x\psi\left(\frac{y}{x}\right)+\frac{1}{2}(x^2+y^2).$$

[This equation, being linear in r, s, t, comes under the head of the case treated in § 85. The student should solve this example by Monge's method, as an exercise.]

Ex. 2. $x^2\frac{\partial^2 z}{\partial x^2}-y^2\frac{\partial^2 z}{\partial y^2}+x\frac{\partial z}{\partial x}-y\frac{\partial z}{\partial y}=0.$

Ex. 3. $x^2\frac{\partial^2 z}{\partial x^2}-4xy\frac{\partial^2 z}{\partial x\,\partial y}+4y^2\frac{\partial^2 z}{\partial y^2}+6y\frac{\partial z}{\partial y}=x^3y^4.$

Other equations may be reducible to linear equations with constant coefficients. (But the transformation is not always so obvious as in the case cited above.) Thus let the student apply the transformation $X=\frac{1}{2}x^2$, $Y=\frac{1}{2}y^2$ to the following example.

Ex. 4. $\frac{1}{x^2}\frac{\partial^2 z}{\partial x^2}-\frac{1}{x^3}\frac{\partial z}{\partial x}=\frac{1}{y^2}\frac{\partial^2 z}{\partial y^2}-\frac{1}{y^3}\frac{\partial z}{\partial y}.$

94. Summary.—The number of classes of partial differential equations of higher order than the first which can be integrated by elementary means is very small. In this chapter we have dealt almost entirely with differential equations either linear in the dependent variable and all of its derivatives, or linear in the highest derivatives only, these being of order two. This latter class frequently yields to Monge's method, § 85.

If the equation is linear in the dependent variable and all of its derivatives, and has constant coefficients, the general method of § 92 applies.

If the linear equation with constant coefficients is "homogeneous," that is, if the dependent variable is absent, and all the derivatives that appear are of the same order, the method of § 88 applies.

If the equation is linear, but the coefficients are not constants, a transformation can sometimes be found to reduce the equation to one with constant coefficients (§ 93).

At times the special method of § 86 can be applied directly to an equation.

Ex. 1. $ys=x+y$.

Ex. 2. $r-s-6t=xy$.

Ex. 3. $zr+p^2=3xy^2$.

Ex. 4. $xr-(x+y)s+yt=\frac{x+y}{x-y}(p-q)$.

Ex. 5. $xr - p = xy.$

Ex. 6. $r - t - 3p + 3q = e^{x+2y}.$

Ex. 7. $x^2r - y^2t = (x + 1)y.$

Ex. 8. $x^2r + 2xys + y^2t + xp + yq - z = 0.$

Ex. 9. $xr + p = 9x^2y^2.$

Ex. 10. $s - t = \frac{x}{y^2}.$

NOTE I

Condition that a Relation exist between Two Functions of Two Variables. — If u and v are two functions of x and y, the necessary and sufficient condition that a relation exist between them is that their *functional determinant* (also called their *Jacobian*) vanishes, that is,

$$\frac{\partial(u, v)}{\partial(x, y)} = J_{x,y}(u, v) = (u, v)_{x,y} = \begin{vmatrix} \dfrac{\partial u}{\partial x} & \dfrac{\partial v}{\partial x} \\ \dfrac{\partial u}{\partial y} & \dfrac{\partial v}{\partial y} \end{vmatrix} = 0.$$

1° To prove the necessity of the condition.

If $u = \phi(v)$, differentiating, we get

$$\frac{\partial u}{\partial x} = \frac{d\phi}{dv}\frac{\partial v}{\partial x},$$

$$\frac{\partial u}{\partial y} = \frac{d\phi}{dv}\frac{\partial v}{\partial y}.$$

These two equations in the single quantity $\frac{d\phi}{dv}$ can hold simultaneously only if

$$\begin{vmatrix} \dfrac{\partial u}{\partial x} & \dfrac{\partial v}{\partial x} \\ \dfrac{\partial u}{\partial y} & \dfrac{\partial v}{\partial y} \end{vmatrix} = 0,$$

which proves the necessity of the condition.

2° Now to prove the sufficiency of the condition.

Suppose u and v to be given by

$$u = f_1(x, y),\ v = f_2(x, y).$$

From these we can eliminate y, resulting in a relation, which may be supposed solved for u, thus

$$u = \phi(x, v).$$

Differentiating, and remembering that x and y are independent variables, we have

$$\frac{\partial u}{\partial x}=\frac{\partial \phi}{\partial x}+\frac{\partial \phi}{\partial v}\frac{\partial v}{\partial x},$$

$$\frac{\partial u}{\partial y}=\qquad\frac{\partial \phi}{\partial v}\frac{\partial v}{\partial y}.$$

If $\begin{vmatrix}\frac{\partial u}{\partial x} & \frac{\partial v}{\partial x}\\ \frac{\partial u}{\partial y} & \frac{\partial v}{\partial y}\end{vmatrix}=0$, these equations can hold simultaneously only provided $\frac{\partial \phi}{\partial x}=0$. But this means that ϕ is free of x. Hence, when the Jacobian vanishes

$$u=\phi(v),$$

which proves the sufficiency of the condition.

Remark.—This theorem can be extended to n functions of n independent variables.

NOTE II

General Summary.—The following is an index to the various methods, given in this book, for solving differential equations:

In the case of a single ordinary differential equation,

if it is of first order and first degree, see § 19;

if it is of first order and higher degree than the first, see § 28 for the general solution, and § 34 for the singular solution;

if it is of higher order than the first and linear with constant coefficients, see § 52 (note what is said there of a very general class of linear equations which can be transformed to linear equations with constant coefficients);

if it is of the second order and linear, see §§ 55, 62, and 74;

if it is of higher order than the first and does not come under any of the above heads, see § 62.

If there is a system of ordinary differential equations, see § 69.

As a final resort, whether there is a single ordinary differential equation or a system of them, the general methods of Chapter XI may be tried.

If there is a single total differential equation in more than two variables, see § 41.

If there is a single partial differential equation of the first order, see § 84.

If there is a single partial differential equation of higher order than the first, see § 94.

The above may be put in the following tabular form:

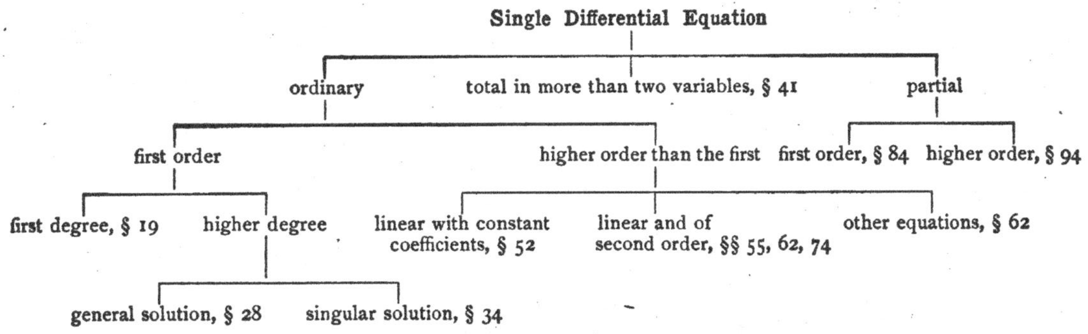

System of Ordinary Differential Equations, § 69

The general methods of Chapter XI apply to single ordinary differential equations and systems of these.

ANSWERS

Section 3

3. $y = x\dfrac{dy}{dx} + \sqrt{1 - \left(\dfrac{dy}{dx}\right)^2}.$
4. $r^2\left(\dfrac{d^2y}{dx^2}\right)^2 = \left[1 + \left(\dfrac{dy}{dx}\right)^2\right]^3.$
5. $x\dfrac{d^2y}{dx^2} - \dfrac{dy}{dx} = 0.$
6. $y = 2x\dfrac{dy}{dx}.$
7. $x\left(\dfrac{dy}{dx}\right)^2 = 2y\dfrac{dy}{dx} + x.$

Section 8

4. $xy + \dfrac{x^2}{2} = c.$
5. $3x^2 - 2xy + x + y^2 - 3y = c.$

Section 9

2. $y + 2xy - 2x^2 = cx^2y.$
3. $y(1 + x^2) = c(1 - y^2).$
4. $\sec x + \tan y = c.$

Section 10

3. $\log x - \dfrac{x}{y} = c.$
4. $y^2 = cx^2(x^2 + y^2).$
5. $x^2 + y^2 = cx^2y^2.$
6. $\log x - \sin\dfrac{y}{x} = c.$

Section 11

2. $\tan^{-1}\left(\dfrac{y+2}{2x+2}\right) + \log[4(x+1)^2 + (y+2)^2] = c.$
3. $5x - 10y + \log(10x + 5y - 2) = c.$

Section 12

2. $x^2y + x^3y^2 = c.$
3. $\log\dfrac{x}{y} - \dfrac{1}{xy} = c.$

Section 13

3. $2y = (x+1)^4 + c(x+1)^2.$
4. $y(1 + x^2)^2 = x^2 + \log x^2 + c.$
5. $y = x^2\left(1 + ce^{\frac{1}{x}}\right).$

Section 14

2. $(y^2 - 1)e^{x^2} = c.$
3. $(\cos y - 1)e^{\cos x} = c.$
4. $x^3y^4(e^x + c) = 1.$
5. $\sqrt{y+1} = x + 1 + c\sqrt{x+1}.$

Section 15

2. $y^2 = cx.$
3. $x^3 + y = cxy^2.$

Section 16

3. $\log\sqrt{x^2 + y^2} - \tan^{-1}\dfrac{y}{x} = c.$
4. $x^2 - y^2 = cx.$
5. $\log x + \dfrac{y^2}{x} = c.$
6. $\tan^{-1}\dfrac{y}{x} = x + c.$

Section 17

1. $3x^4 + 8x^3y + 6x^2y^2 = c.$
2. $(x^2 + y^2)e^y = c.$
3. $xy + y^2 + \dfrac{2x}{y^2} = c.$

4. $y = cx.$

6. $x^2 + y^2 = c(my - x).$

Section 18

4. $y = \sqrt{\frac{c}{b}}\,x^a \dfrac{ke^{\frac{2\sqrt{bc}}{a}x^a} - 1}{ke^{\frac{2\sqrt{bc}}{a}x^a} + 1},$

if b and c have the same sign;

$= \sqrt{-\frac{c}{b}}\,x^a \tan\left(k - \frac{\sqrt{-bc}}{a}x^a\right),$

if b and c have opposite signs.

Section 19

1. $\sqrt{1 - x^2} + \sqrt{1 - y^2} = c.$
2. $x\sqrt{1 - y^2} + y\sqrt{1 - x^2} = c.$
3. $y + x^3 + 3 = ce^{\frac{x^3}{3}}.$
4. $\dfrac{y - x + 1}{y - x - 1} = ce^{2y}.$
5. $x^3y^3(3e^x + c) = 1.$
6. $xy = ce^{x+y}.$
7. $\log y + \dfrac{x}{y} = c.$
8. $c^2x^2 - 2cy - 1 = 0.$
9. $y = (x + a)^5 + c(x + a)^3.$
10. $\sin^{-1}\dfrac{y}{x} = \log x + c.$
11. $x \sin\dfrac{y}{x} = c.$
12. $(x + y - 1)^3 = c(x - y + 3).$
13. $y = \dfrac{x}{\sqrt{1 - x^2}} + ce^{-\frac{x}{\sqrt{1-x^2}}}.$
14. $\dfrac{1}{y} = c\sqrt{1 - x^2} - a.$
15. $x^3y^2 - x^2 = cy.$
16. $y = \tan^{-1} x - 1 + ce^{-\tan^{-1}x}.$
17. $x^{\frac{5}{2}}y^{\frac{3}{2}} - x^{\frac{3}{2}}y^{\frac{7}{2}} = c.$
18. $y = \sin x - 1 + ce^{-\sin x}.$
19. $\dfrac{x^2}{2} + \dfrac{x}{y} = c.$
20. $\dfrac{x}{y} = ce^{x+y}.$
21. $\log x^3 + \log y + \dfrac{y^2}{2} = c.$
22. $x^2 + y^2 = c\sqrt{x^2 + y^2}\,e^{\tan^{-1}\frac{y}{x}} + \frac{1}{2}(y - x).$
23. $x + y - 4\log(2x + 3y + 7) = c.$
24. $x^2y^2(y^2 - x^2) = c.$
25. $x^2 + y^2 + \dfrac{1}{xy} = c.$
26. $\sqrt{1 + x^2 + y^2} + \tan^{-1}\dfrac{x}{y} = c.$
27. $x + ye^{\frac{x}{y}} = c.$
28. $\dfrac{1}{y} = \log x + 1 + cx.$
29. $x^2y^2 - 2xy \log cy = 1.$
30. $y = ce^{-\sqrt{\frac{x}{y}}}.$

Section 20

3. $xy\left(\dfrac{dy}{dx}\right)^2 + (x^2 - y^2 - \lambda)\dfrac{dy}{dx} - xy = 0.$
4. $y\left(\dfrac{dy}{dx}\right)^2 + 2x\dfrac{dy}{dx} - y = 0.$
5. $y = x\dfrac{dy}{dx} + r\sqrt{1 + \left(\dfrac{dy}{dx}\right)^2}.$
6. $x\left(\dfrac{dy}{dx}\right)^2 + (\lambda - x - y)\dfrac{dy}{dx} + y = 0.$
7. $2x\left(\dfrac{dy}{dx}\right)^2 = (3x - 1)^2.$
8. $\left(2x\dfrac{dy}{dx} - y\right)^2 = 8x^3.$
9. $x^2\left(\dfrac{d^2y}{dx^2}\right)^2 - 2x\dfrac{dy}{dx}\dfrac{d^2y}{dx^2}\left[1 + \left(\dfrac{dy}{dx}\right)^2\right]$

$= \left[1 + \left(\dfrac{dy}{dx}\right)^2\right]^2.$
10. $y\dfrac{dy}{dx} - x\left(\dfrac{dy}{dx}\right)^2 - xy\dfrac{d^2y}{dx^2} = 0.$
11. (a) $3\left(\dfrac{d^2y}{dx^2}\right)^2 = \dfrac{dy}{dx}\dfrac{d^3y}{dx^3},$ (b) $\dfrac{d^3y}{dx^3} = 0.$
12. $\left[1 + \left(\dfrac{dy}{dx}\right)^2\right]^3 = r^2\left(\dfrac{d^2y}{dx^2}\right)^2.$

Section 21

6. The parabolas $y^2 = 2\lambda x + c$.
7. $y = ce^{-\frac{1}{x}}$.
8. The circles $x^2 + y^2 = cx$.
9. The cardioids $\rho = c(1 - \cos\theta)$.
10. The spirals $\rho^2 = ce^{\pm\theta}$.

Section 22

6. The circles $x^2 + y^2 = c\,(my - x)$, (see Ex. 6, § 17), through the origin with their centers on the line $y = -mx$, where $m = \tan\alpha$.
7. The logarithmic spirals $\rho e^{m\theta} = c$ (compare Ex. 5).
8. The ellipses $2x^2 + y^2 = c$.
9. The equilateral hyperbolas $xy = c$.
10. $y^a = cx^b$.
12. $\rho = c(1 + \cos\theta)$.
13. $\rho = e^{\sqrt{c-\theta^2}}$.
14. $\rho^m \cos m\theta = c^m$.
15. $\tan\theta = ce^{2\rho}$.
16. $\rho = \dfrac{2c}{1 + \cos\theta}$, the family of parabolas, confocal and coaxial with the original family (compare Ex. 4).
17. $\rho = \dfrac{\lambda^2 - c^2}{\lambda\cos\theta - c}$, the same family of conics (compare Ex. 3).

Section 23

2. $v = gt\sin\alpha$; $x - x_0 = \frac{1}{2}gt^2\sin\alpha$.
3. $x - x_0 = \dfrac{g}{r^2}\log\dfrac{ce^{rt} + e^{-rt}}{c + 1}$.
5. $x - x_0 = \dfrac{v_0}{k}(e^{kt} - 1)$. If $k > 0$, the velocity and the distance covered increase indefinitely; if $k < 0$, the velocity diminishes continually, but never vanishes at a finite time, while the distance covered increases continually, but never attains the value $-\dfrac{v_0}{k}$ at a finite time.
6. (a) $i = ce^{-\frac{Rt}{L}}$, (b) $i = \dfrac{E}{R} + ce^{-\frac{Rt}{L}}$,

$$(d)\ i = ce^{-\frac{Rt}{L}} + \frac{e^{-\frac{Rt}{L}}}{L}\int E(t)\,e^{\frac{Rt}{L}}\,dt.$$

Section 24

4. $y^2 = 2x(x - c)^2$.
5. $2x = ce^y - \dfrac{1}{ce^y}$.
6. $(1 + xy - cy)(x + y + 1 - ce^x)(x - y - 1 - ce^{-x}) = 0$.

Section 25

4. $y = x + \dfrac{c + e^{2x}}{c - e^{2x}}$.
5. $xy = c^2x + c$.
6. $4(x^2 + y)^3 = (2x^3 + 3xy + c)^2$.

Section 26

1. $y = \dfrac{c}{(1 + p^2)^{\frac{3}{2}}}$, $x = \dfrac{-cp(2p^2 + 3)}{(1 + p^2)^{\frac{3}{2}}}$
2. $y^2 - 2cx + a^2c^2 = 0$.
4. $y = c(x - c)^2$.
5. The family of circles, $(x - c)^2 + y^2 = c$

Section 27

4. $e^y = c(e^x + 1) + c^3$.
5. $y^2 = cx^2 + \dfrac{1}{c}$.
6. $x^2 + y^2 = c(x + y) - \dfrac{c^2}{4}$.
7. $y^2 = cx + \dfrac{1}{8}c^3$.
9. $y = cx^2 + c^2x$.

Section 28

1. $(x-c)^2+y^2=a^2$.
2. $cy=c^2(x-b)+a$.
3. $x+cxy+c^2=0$.
4. $x^2+c(x-3y)+c^2=0$.
5. $(x-y+1)^2+2c(x+y+1)+c^2=0$.
6. $y^2=cx^2-\dfrac{a^2c}{c+1}$.
7. $y(1\pm\cos x)=c$.
8. $(y-cx)^2=1-c^2$.
9. $(y-cx)^2=4c^2$.
10. $y^2=cx^2+c^2$.
11. $x+2cye^x=c^2xe^{2x}$.
12. $\theta=\displaystyle\int\frac{f(\rho)d\rho}{\rho\sqrt{\rho^2-[f(\rho)]^2}}+c$.
13. $\rho=ce^{\theta\frac{\sqrt{1-k^2}}{k}}$.
14. $x=k\left(\tan^{-1}p+\dfrac{p}{1+p^2}\right)+c$,
    $y=\dfrac{kp^2}{1+p^2}$; or putting $p=\tan\dfrac{\theta}{2}$,
    $x=\dfrac{k}{2}(\theta+\sin\theta)+c, y=\dfrac{k}{2}(1-\cos\theta)$,
    a family of cycloids generated by a circle of radius $\dfrac{k}{2}$. We see, then, that a characteristic property of a cycloid generated by a circle of radius a is that $s^2=8\,ay$, where s is the length of arc measured from the nearest cusp.
15. The same family of cycloids as in Ex. 14. Here $k=\dfrac{1}{2}gt^2$.

Section 29

1. $y^2=r^2$.

Section 32

1. $\dfrac{x^2}{c^2+k}+\dfrac{y^2}{k}=1$, when the fixed points are $(c, 0)$, $(-c, 0)$, and k is the constant product. According as k is positive or negative, we have the ellipse or the hyperbola, having the fixed points for foci, and $\sqrt{\pm k}$ for semi-conjugate axis.
2. The circle $x^2+y^2=k^2$, where k is the constant distance.
3. The equilateral hyperbola $2\,xy=a^2$.
4. The parabola $(x-y)^2-2\,a(x+y)+a^2=0$.
5. $(y-cx)^2+4c=0$, g. s.*,
   $xy-1=0$, s. s.
   $(y-cx)^2=b^2+a^2c^2$, g. s.,
   $\dfrac{x^2}{a^2}+\dfrac{y^2}{b^2}=1$, s. s.,
   $(y+x-c)^2=4\,xy$, g. s.,
   $xy=0$, s. s.

Section 33

4. $(x-c)^2=y^3-y^4$, g. s., $y=1$, s. s., $y=0$, c. l., $y=\frac{3}{4}$, t. l.
   § 24, Ex. 3, $y^2-1=0$, s. s.
   Ex. 4, there is no s. s., $x=0$, t.l. and p.s. for $c=\infty$, $y=0$, n.l.
   Ex. 5, $x^2+1=0$, s. s.
   § 25, Ex. 5, $4\,x^2y+1=0$, s. s., $x=0$, t. l.
   Ex. 6, there is no s. s. $x^2+y=0$, c. l.
   § 26, Ex. 2, $x^2-a^2y^2=0$, s. s.
   Ex. 4, $4\,x^3-27\,y=0$, s. s., $y=0$, s.s. and t.l.
   § 27, Ex. 2, $x^2+4\,e^{2y}=0$, s. s.
   Ex. 6, $xy=0$, s. s., $y-x=0$, t. l.
   § 28, Ex. 2, $y^2-4\,a(x-b)=0$, s. s.
   Ex. 3, $xy^2-4=0$, s. s., $x=0$, p. s. for $c=0$.

* In these answers the following abbreviations are used: g. s. for general solution, s. s. for singular solution, p. s. for particular solution, t. l. for tac-locus, n. l. for nodal locus, c. l. for cuspidal locus.

Ex. 4, $x^2 + 2xy - 3y^2 = 0$, s.s.
Ex. 5, $(x + 1)y = 0$, s. s.
Ex. 11, $x^2 + y^2 = 0$, s. s., $x = 0$, t. l.

5. $4y^2 = 4x + 1$, s. s., $y = 0$, t. l.

Section 36

2. $z = ce^{\frac{z}{y}}$.
3. $x^2 + y^2 + (z - c)^2 = a^2$.
4. $x^2 + y^2 + z^2 = cx$.

Section 37

2. $xy + yz + zx = c(x + y + z)$.
3. $\dfrac{y + z}{x} + \dfrac{z + x}{y} = c$.

Section 41

1. $yz + zx + xy = c$.
2. $x^2 + y^2 = c(z + 1)^2$.
3. $e^x(x + y^2 + z^2) = c$.
4. $x = \dfrac{z}{y + a} + c$.
5. $e^x(y + z) = c$.
6. $(x^2 + y + z)e^{x^2} = c$.
7. $x^2 + xy^2 + x^2z - t = c$.
8. $z = ce^{-\frac{y}{x}}$.
9. $(x - 1)(y - 1)(z - 1)e^{x+y+z} = c$.
10. $(z - y)^2(x + 2y) = c$.
11. $(y + z)e^{x+y} = ct$.
12. $(y + z)(t + c) + z(x - t) = 0$.

Section 43

3. $y = c_1 + c_2e^x + c_3e^{-x}$.
4. $y = c_1e^x + c_2e^{-x} + c_3e^{2x}$.

Section 44

2. $y = e^x(c_1 + c_2x) + c_3e^{-x}$.
3. $y = e^{-x}(c_1 + c_2x + c_3x^2) + c_4e^x$.
4. $y = c_1 + e^{3x}(c_2 + c_3x)$.

Section 45

2. $y = (c_1 + c_2x)\cos x + (c_3 + c_4x)\sin x$.
3. $y = c_1 + e^{\frac{1}{2}x}\left(c_2 \cos \frac{x}{2}\sqrt{3} + c_3 \sin \frac{x}{2}\sqrt{3}\right)$.

Section 47

2. $y = c_1e^{-x} + c_2e^{-2x} + e^{-2x}e^{e^x}$.
3. $y = \left(c_1 + c_2x + c_3x^2 + \dfrac{20x^3 - x^5}{60}\right)e^{-x}$.
4. $y = c_1e^{-x} + c_2e^{2x} + \dfrac{\cos x - 3\sin x}{10}$.
5. $y = (c_1 + c_2x)e^x - e^x \log(1 - x)$.

Section 48

1. $y = c_1e^x + c_2e^{2x} - xe^x$.
2. $y = c_1e^x + c_2e^{-x} + c_3e^{3x} + \dfrac{9x^2 + 6x + 20}{27}$.
3. $y = c_1 \cos x + c_2 \sin x + x \sin x + \cos x \log \cos x$.
4. $y = (c_1 + c_2x)e^x + c_3e^{2x} - \dfrac{x}{2} - \dfrac{5}{4}$.

Section 49

2. $y = c_1 \cos x + c_2 \sin x - \cos x \log(\sec x + \tan x)$.

Section 50

3. $y = c_1 \cos x + c_2 \sin x + e^x + x^3 - 7x$.
4. $y = (c_1 + c_2x)e^{-x} + \frac{1}{3}e^{2x} - \frac{1}{2}\sin x$.
5. $y = c_1e^x + e^{-\frac{1}{2}x}\left(c_2 \cos \frac{x}{2}\sqrt{3} + c_3 \sin \frac{x}{2}\sqrt{3}\right) - x^2$.
6. $y = c_1 + c_2e^{-x} + c_3e^{3x} - \frac{1}{3}x^3 + \frac{2}{3}x^2 - \frac{14}{9}x + \frac{1}{10}\sin x + \frac{1}{5}\cos x$.

7. $y=(c_1+c_2x)e^x+(c_3+c_4x)e^{-x}+\frac{1}{8}x^2e^x+4.$

8. $y=c_1+c_2e^{2x}+\frac{x}{2}(e^{2x}-1).$

9. $y=(c_1+c_2x)\cos x+(c_3+c_4x)\sin x-\frac{1}{8}x^2\cos x.$

Section 51

2. $y=(c_1\cos\log x+c_2\sin\log x)x+\frac{c_3}{x}+5x+\frac{2\log x}{x}.$

3. $y=\frac{c_1+c_2\log x}{x}+\frac{1}{x}\log\frac{x}{1-x}.$

4. $y=c_1(x+1)^2+c_2(x+1)^3+\frac{3x+2}{6}.$

Section 52

1. $y=c_1e^{2x}+c_2e^{3x}+\frac{\cos x-\sin x}{10}+xe^{2x}.$

2. $y=c_1e^x+c_2e^{-x}+c_3\sin x+c_4\cos x-\frac{1}{5}e^x\cos x.$

3. $y=(c_1+c_2x)e^{-x}+2x^3-12x^2+36x-48-\frac{(2x-1)e^{3x}}{32}.$

4. $y=(c_1+c_2x+c_3x^2)e^{-x}+\frac{x^4e^{-x}}{24}.$

5. $y=c_1+c_2e^{2x}+c_3e^{-2x}-\frac{x^3}{12}-\frac{x}{8}-\frac{3}{8}xe^{2x}.$

6. $y=(c_1+c_2x)e^x+(c_3+c_4x)e^{-x}+\frac{1}{4}\cos x.$

7. $y=(c_1+c_2\log x)\sin\log x+(c_3+c_4\log x)\cos\log x+(\log x)^2+2\log x-3.$

8. $y=c_1+(c_2+c_3x)e^{-x}+\frac{x^3}{3}-\frac{5x^2}{2}+8x.$

9. $y=c_1\sin 2x+c_2\cos 2x+\frac{1-x\sin 2x}{8}.$

10. $y=a\cos(x+b)+\sin x\log(\tan x+\sec x)-1.$

11. $y=(c_1+c_2x+c_3x^2)e^x-x-3-\frac{x^6e^x}{120}.$

12. $y=(c_1+c_2x+c_3x^2)e^x+c_4e^{-2x}+\frac{1}{40}e^{3x}.$

13. $y=c_1\cos x+c_2\sin x+\frac{x^2\sin x+x\cos x}{4}.$

14. $y=x(c_1+c_2\log x)+\frac{c_3}{x}+\frac{\log x}{4x}.$

15. $y=c_1e^x+e^{-\frac{x}{2}}\left(c_2\cos\frac{x}{2}\sqrt{3}+c_3\sin\frac{x}{2}\sqrt{3}\right)+\frac{xe^x}{6}(x-2)-\frac{\cos 2x+8\sin 2x}{130}-\frac{1}{2}.$

16. $y=(c_1+c_2x)e^x-\frac{1}{2}\sin x+e^x\left(\frac{x^3}{2}+\frac{x^4}{12}\right).$

18. 1°. $k>n.$ $\theta=e^{-kt}(Ae^{\mu t}+Be^{-\mu t})$, where $\mu^2=k^2-n^2$. θ diminishes continually but vanishes only for $t=\infty$, theoretically. Practically, the swing of the pendulum is soon damped sufficiently, so that the assumed law ceases to hold, and θ becomes zero in a finite time. But the solution tells us that the pendulum comes to rest without vibrating.

2°. $k<n.$ $\theta=Ae^{-kt}\cos(\mu t+B)$, where $\mu^2=n^2-k^2$. The pendulum vibrates with constant period $\frac{2\pi}{\sqrt{n^2-k^2}}$, which is longer than

in the case of motion in a vacuum (Ex. 17). The amplitude is Ae^{-kt}, which diminishes continually.

3°. $k = n$. $\theta = (A + Bt)e^{-kt}$. If A and B have the same sign, θ diminishes continually, as in case 1°. If A and B have opposite signs, θ passes through zero, changes sign, attains a maximum in absolute value and then diminishes continually in absolute value, as before.

19. (a) 1°. $m \neq n$. $\theta = A\cos(nt+B) + \dfrac{C}{n^2-m^2}\cos mt$.

2°. $m = n$. $\theta = A\cos(nt+B) + \dfrac{Ct}{2n}\sin nt$.

(b) The same complementary function as in Ex. 18*.

$$+\frac{C[(n^2-m^2)\cos mt + 2k\sin mt]}{(n^2-m^2)^2+4k^2}.$$

This last term may also be written

$$\frac{C}{\sqrt{(n^2-m^2)^2+4k^2}}\cos(mt-\alpha),$$

where $\tan\alpha = \dfrac{2k}{n^2-m^2}$. The part of the motion indicated by this term is called the forced vibration. Its period is $\dfrac{2\pi}{m}$. It is not in phase with the periodic force [except in case (a) 1°], lagging behind by the angle α. When $n > m$, $0 < \alpha < \dfrac{\pi}{2}$; when $n < m$, $\dfrac{\pi}{2} < \alpha < \pi$; when $n = m$, $\alpha = \dfrac{\pi}{2}$. In this case the forced vibration is given by $U = \dfrac{C}{2k}\sin nt$ for case (b), and by $U = \dfrac{Ct}{2n}\sin nt$ for case (a). If k is small, the amplitude is large in case (b), while in case (a) the amplitude increases with the time. This explains resonance in Acoustics, the effect of the measured step of soldiers on a bridge, and the like.

20. $x = x_0\cos kt + \dfrac{v_0}{k}\sin kt$. The period is independent of x_0 and v_0.

21. $2x = \left(x_0 + \dfrac{v_0}{k}\right)e^{kt} + \left(x_0 - \dfrac{v_0}{k}\right)e^{-kt}$.

22. $r = c_1e^{\omega t} + c_2e^{-\omega t} + \dfrac{g}{2\omega^2}\sin\omega t$, where

$$c_1 + c_2 = r_0,\quad \omega(c_1 - c_2) + \frac{g}{2\omega} = v_0.$$

If $c_1 \neq 0$, the first term soon predominates, and the motion is spiral.

If $c_1 = 0$, the second term soon becoming negligible, simple harmonic motion results.

When $r_0 = 0$ and $v_0 = \dfrac{g}{2\omega}$, $r = \dfrac{g}{2\omega^2}\sin\omega t$; *i.e.* we have simple harmonic motion from the beginning.

Section 53

2. $y = c_1e^x + c_2x^2e^x + x$.

3. $y = c_1x + c_2(1 + x\tan^{-1}x)$.

4. $y = c_1x + c_2e^x + x^2 + 1$.

6. $y = \sec x(c_1e^{ax} + c_2e^{-ax})$.

7. $y = x^{\frac{1}{2}}e^{-\frac{x^2}{4}}(c_1 + c_2\log x)$.

8. $xy = c_1e^x + c_2e^{-x} + xe^x$.

Section 54

2. $y = c_1x\sqrt{1-x^2} + c_2(1 - 2x^2)$.

3. $y = c_1\sin(\sin x) + c_2\cos(\sin x)$.

4. $y = c_1\cos\dfrac{1}{2x^2} + c_2\sin\dfrac{1}{2x^2} + \dfrac{1}{x^2}$.

5. $y = c_1e^{2x^2} + c_2e^{-x^2} - \dfrac{1}{3}x^2e^{-x^2}$.

* Replace k by km throughout the answer.

Section 55

1. $y = c_1e^x + c_2(x^3 + 3x^2 + 6x + 6)$.
2. $y = c_1e^x + c_2e^{3x}(4x^3 - 42x^2 + 150x - 183)$.
3. $y = \frac{(c_1e^x + c_2e^{-x})}{x^2}$.
4. $y = c_1(x^2 - 1) + c_2x$.
5. $y = e^x(c_1 + c_2 \log x)$.
6. $y = x^2(c_1 \cos x + c_2 \sin x)$.
7. $y = c_1x + c_2(x^3 + 1)$.
8. $2y = x(c_1e^{2x} + c_2 - x)$.
9. $y = c_1x^n \cos(ax + c_2)$.
10. $y = c_1 \cos \frac{n}{x} + c_2 \sin \frac{n}{x}$.

Section 57

3. $y = x + c_1 \int e^{-\frac{x^2}{2}} dx + c_2$.
4. $y = (x - 2)e^x + c_1x + c_2$.
5. $y = \frac{1}{2}c_1x^2 + x\sqrt{1 + c_1^2} + c_2$.

Section 58

2. $(x - a)^2 + y^2 = b$.
3. $e^y = \frac{4ac^2e^{cx}}{(1 - ae^{cx})^2}$.
4. $ay + be^{ax} + 2 = 0$.

Section 60

2. $xy = \frac{x^2}{4} + c_1 \log x + c_2$.
3. $(x - 1)^2y = c_1 + c_2x - \cos x$.
4. $xy\sqrt{x^2 - 1} = c_1\sqrt{x^2 - 1} + c_2 \log(x + \sqrt{x^2 - 1}) + c_3$.
5. $x^3y^2 = c_1x^2 + c_2x + c_3$.
7. $xy = c_1x - c_1 \sin^{-1}x \sqrt{1 - x^2} + c_2\sqrt{1 - x^2}$.
8. $y = c_1 \cos x^2 + c_2 \sin x^2 + c_3x^2$.
10. $\tan y = c_1 \cot x + c_2$.

Section 61

2. $y = x \log \frac{c_1x}{1 + c_2x}$.
3. $\log y = c_1e^x + c_2e^{-x} + x^2 + 2$.
4. $y = c_1 \cot x + c_2(1 - x \cot x)$.

Section 62

1. $y = c_1 - \log \cos(x + c_2)$.
2. $y = (\sin^{-1} x)^2 + c_1 \sin^{-1} x + c_2$.
3. $\frac{a + y}{a - y} = e^{a(x+b)}$.
4. $(1 + x^3)y = c_1x^2 + c_2x + c_3$.
5. $(x - 1)^5y = c_1\left(x^4 - 6x^2 + 2x - \frac{1}{3} - 4x^3 \log x\right) + c_2x^3$.
6. $\log y = 1 + \frac{1}{c_1x + c_2}$.
7. $y = c_1 \log x + c_2$.
8. $x^2y + xy^2 = c_1x + c_2 - \frac{x^4}{12}$.
9. $y = \log \cos(c_1 - x) + c_2$.
10. $y = \frac{x^2}{2} + c_1\sqrt{x^2 - 1} + c_2$.
11. $y = c_1 + (c_2 + c_3 \log x)\sqrt{x}$.
12. $y = c_1 \sin^2 x + c_2 \cos x - c_2 \sin^2 x \log \tan \frac{x}{2}$.
13. (*a*) The circles $(x + c_1)^2 + y^2 = c_2$.
    (*b*) The catenaries
    $$y = \frac{c_1}{2}\left(e^{\frac{x+c_2}{c_1}} + e^{-\frac{x+c_2}{c_1}}\right).$$
14. (*a*) The cycloids
    $$x + c_1 = c_2 \operatorname{vers}^{-1}\frac{y}{c_2} - \sqrt{2c_2y - y^2},$$
    (*b*) the parabolas
    $$(x + c_1)^2 = 2c_2y - c_2^2.$$
15. The central conics
    $$c_1y^2 - \frac{c_1^2}{k}(x + c_2)^2 = 1;$$ hyperbolas when $k > 0$, ellipses when $k < 0$.

16. $y = \frac{1}{2}\left\{\frac{c_1(a-x)^{1-\frac{1}{n}}}{\frac{1}{n}-1} + \frac{1}{c_1}\frac{(a-x)^{1+\frac{1}{n}}}{1+\frac{1}{n}}\right\} + c_2$, if $n \neq 1$.

$y = \frac{(a-x)^2}{4c_1} - \frac{c_1}{2}\log(a-x) + c_2$,

if $n = 1$.

17. $v \equiv l\frac{d\theta}{dt} = \sqrt{2gl(\cos\theta - \cos\alpha)}$.

18. (*a*) $v^2 = 2k^2\left(\frac{1}{x} - \frac{1}{a}\right)$,

(*b*) $t = \frac{1}{k}\sqrt{\frac{a}{2}}\left(\sqrt{ax - x^2} - \frac{a}{2}\text{vers}^{-1}\frac{2x}{a} + \frac{a\pi}{2}\right)$,

(*c*) $\frac{a}{2}$,

(*d*) $v^2 = 2gR^2\left(\frac{1}{R} - \frac{1}{R+h}\right)$;

if $h = \infty$, $v = \sqrt{2gR}$, which is 7 miles per second, approximately.

Section 64

1. $x = c_1e^{\frac{1}{3}t} + c_2e^{-\frac{1}{3}t} - 6t$,

$y = -2c_1e^{\frac{1}{3}t} - c_2e^{-\frac{1}{3}t} + \frac{1}{2}e^t + 9t + 9$.

2. $x = (6c_2 - 2c_1 - 2c_2t)e^t - \frac{1}{3}c_3e^{-\frac{3}{2}t} - \frac{1}{3}$,

$y = (c_1 + c_2t)e^t + c_3e^{-\frac{3}{2}t} - \frac{1}{2}t$.

3. $x = (c_1 + c_2t)e^t + (c_3 + c_4t)e^{-t}$,

$2y = (c_2 - c_1 - c_2t)e^t - (c_3 + c_4 + c_4t)e^{-t}$.

Section 65

6. $x^3 - y^3 = c_1$, $\quad x^2 - t^2 = c_2$.

7. $x^2 + y^2 = c_1$, $\quad t = \frac{c_2x + y}{x - c_2y}$.

8. $x^2 - y^2 = c_1$, $\quad 2(x+y) - t^2 = c_2$.

9. $y = c_1t$, $\quad x^2 + y^2 + t^2 = c_2t$.

10. $x - y = c_1(x - t)$,

$(x + y + t)(y - t)^2 = c_2$.

11. $x - y - t = c_1$, $\quad x^2 + y^2 - c_2t^2 = 0$.

12. $x^3y^3t = c_1$, $\quad \frac{y}{x^2} + \frac{x}{y^2} = c_2$.

Section 66

$x^2 + z^2 = c_1$, $y^2 + z^2 = c_2$.

Section 69

1. The path is a parabola lying in the vertical plane determined by the direction of the initial velocity. Taking the initial position for origin, the horizontal line of the plane through it for the x axis, and the vertical line through it for the y axis, the equations of the path are

$$x = v_0t\cos\alpha,\ y = v_0t\sin\alpha - \frac{1}{2}gt^2.$$

Or, eliminating t, we have

$$y = x\tan\alpha - \frac{gx^2}{2v_0^2\cos^2\alpha}.$$

2. $x = \frac{v_0\cos\alpha}{k}(1 - e^{-kt})$,

$y = \frac{kv_0\sin\alpha + g}{k^2}(1 - e^{-kt}) - \frac{gt}{k}$.

3. $x = a\cos kt + \frac{v_0\cos\alpha}{k}\sin kt$,

$y = \frac{v_0\sin\alpha}{k}\sin kt$.

Or, eliminating t, we have

$$(x\sin\alpha - y\cos\alpha)^2 + \frac{a^2k^2y^2}{v_0^2} = a^2\sin^2\alpha,$$

an ellipse with its center at the origin. Since x and y are periodic functions of period $\frac{2\pi}{k}$, we see that the motion is periodic, the period being independent of the dimensions of the ellipse.

4. $2kx = (ka + v_0 \cos\alpha)e^{kt} + (ka - v_0\cos\alpha)e^{-kt}$,
$2ky = v_0 \sin\alpha\,(e^{kt} - e^{-kt})$.
Or, eliminating t, we have
$$(x\sin\alpha - y\cos\alpha)^2 - \frac{a^2k^2y^2}{v_0^2} = a^2\sin^2\alpha,$$
a hyperbola with its center at the origin.

5. $$\left.\begin{aligned} p &= a\sin(kt+b),\\ q &= a\cos(kt+b),\\ r &= c,\end{aligned}\right\}$$
where $a = \sqrt{p_0^2 + q_0^2}$,
$b = \tan^{-1}\frac{p_0}{q_0}$, $c = r_0$, $k = r_0\frac{A-C}{A}$.
Angular velocity
$\equiv \omega = \sqrt{p_0^2 + q_0^2 + r_0^2}$, a constant.
Direction cosines of the instantaneous axis of rotation are $\frac{p}{\omega}, \frac{q}{\omega}, \frac{r_0}{\omega}$.

6. The paths are the curves of intersection of the hyperbolic cylinders
$$\frac{x^2}{a^2} - \frac{z^2}{c^2} = c_1,\quad \frac{y^2}{b^2} - \frac{z^2}{c^2} = c_2.$$
The time is given by
$$t - t_0 = \frac{\pm 1}{ab}\int_{z_0}^{z}\frac{dz}{\sqrt{(z^2+c^2c_1)(z^2+c^2c_2)}}.$$
For the curve through the origin $c_1 = c_2 = 0$. Hence the path is one of the lines
$$\frac{x}{a} = \pm\frac{y}{b} = \pm\frac{z}{c}.$$
The time in covering a part of it is
$$t - t_0 = \frac{\pm 1}{ab}\left(\frac{1}{z_0} - \frac{1}{z}\right) = \frac{\pm 1}{ca}\left(\frac{1}{y_0} - \frac{1}{y}\right) = \frac{\pm 1}{bc}\left(\frac{1}{x_0} - \frac{1}{x}\right).$$

Section 74

3. $y = A\left(1 - \frac{1}{2}x + \frac{1}{10}x^2\right) + B(x^{-3} - 5x^{-2} + 10x^{-1} - 10 + 5x - x^2)$,
or $y = A(x^2 - 5x + 10 - 10x^{-1} + 5x^{-2} - x^{-3}) + B\left(x^{-1} - \frac{1}{2}x^{-2} + \frac{1}{10}x^{-3}\right)$.

4. $$y = A\left(x + \frac{x^3}{2\cdot 5} + \frac{x^5}{2\cdot 4\cdot 5\cdot 9} + \frac{x^7}{2\cdot 4\cdot 6\cdot 5\cdot 9\cdot 13} + \cdots\right)$$
$$+ Bx^{\frac{1}{2}}\left(1 + \frac{x^2}{2\cdot 3} + \frac{x^4}{2\cdot 4\cdot 3\cdot 7} + \frac{x^6}{2\cdot 4\cdot 6\cdot 3\cdot 7\cdot 11} + \cdots\right)$$
$$+ \frac{x^2}{1\cdot 3} + \frac{x^4}{1\cdot 3\cdot 3\cdot 7} + \frac{x^6}{1\cdot 3\cdot 5\cdot 3\cdot 7\cdot 11} + \cdots.$$

5. $y = A(6 - 4x + x^2) + B(x^{-2} - 4x^{-1})$.

Section 75

5. $F(1, 1, 1, x)$,
$2F\left(-\frac{n}{2}, -\frac{n-1}{2}, \frac{1}{2}, x^2\right)$,
$2nxF\left(-\frac{n-1}{2}, -\frac{n}{2}+1, \frac{3}{2}, x^2\right)$,
$\text{Limit}_{\alpha,\beta=\infty} F\left(\alpha, \beta, \frac{1}{2}, -\frac{x^2}{4\alpha\beta}\right)$,
$\text{Limit}_{\alpha,\beta=\infty} 2F\left(\alpha, \beta, \frac{1}{2}, \frac{x^2}{4\alpha\beta}\right)$.

Section 76

1. $x - y + z(p - q) = 0$.
2. $xq - yp = 0$.
3. $z = px + qy + \sqrt{p^2 + q^2}$.
4. $z = pq$.
5. $(1 + p^2)^2(p^2 + q^2 + 1) = r^2$,
$(1 + q^2)^2(p^2 + q^2 + 1) = t^2$,
$p^2q^2(p^2 + q^2 + 1) = s^2$.
6. $z = xp + yq - xys$, $r = 0$, $t = 0$.
7. $r = 0$, $s = 0$, $t = 0$.

Section 77

2. $xzp + yzq = xy$.

3. $yp - xq = y^2 - x^2$.
4. $r - t = 0$.
5. $xr - (x+y)s + yt = \frac{x+y}{x-y}(p-q)$.

Section 79

3. $\phi(xy+z,\ x^2+y^2) = 0$.
4. $\phi(x^2+y^2+z^2,\ x+y+z) = 0$.

Section 82

2. $y = (b-x)(a+yz)$.*
3. $6z = (2x-a)^3 + 6a^2y + b$.

Section 83

3. $e^z = cx^a y^{1-a}$. Since the equation is linear its general solution is
$$e^z = xf\left(\frac{x}{y}\right).$$
4. $z = cx^a y^{\sqrt{1-a^2}}$.
5. $z = ax + a\log y + c$. Since the equation is linear, its general solution is $z = f(x + \log y)$.
7. $az - 1 = cxe^{ay}$.
8. $4az = (x + ay + b)^2$.
9. $z = ax + a^2y^2 + b$.
10. $\log z = (x+a)^2 + (y+a)^2 + b$.
11. $z = ax + by + \sqrt{a^2+b^2+1}$. Singular solution is $x^2+y^2+z^2 = 1$.
12. $\frac{z}{a} = \log(x+y-1) + y + b$. Since the equation is linear, its general solution is
$\log(x+y-1) + y = f(z)$.
13. $z = c(x+y+a)e^{ay}$.
15. $2z = x^2\phi(y) + \frac{y}{\phi(y)}$.

* Attention should be called to the fact that no unique answer can be given for the complete solution. Other forms just as good as the ones here given may be found by selecting various forms for the auxiliary function ϕ.

16. $z = x - \frac{1}{y} + e^{-xy}\phi(y)$.
17. $z = y\sin^{-1}\frac{x}{y} + \phi(y)$.

Section 84

1. $\phi\left(\frac{1}{x} - \frac{1}{y},\ \frac{1}{y} - \frac{1}{z}\right) = 0$.
2. $z = ax + (a^2+1)y + b$.
3. $xyz - 3u = \phi\left(\frac{y}{x},\ \frac{z}{x}\right)$.
4. $z = ax + by + (a+b)^2$.
5. $z = cx^a y^{\frac{1}{a}}$.
6. $x^2 - z^2 = \phi(x^3 - y^3)$.
7. $z = axe^y + \frac{1}{2}a^2e^{2y} + b$.
8. $\sqrt{1+a}\,z = 2\sqrt{x+ay} + b$.
9. $z = a\sqrt{x+y} + \sqrt{1-a^2}\sqrt{x-y} + b$.
10. $z^2 - a^2x^2 = (ay+b)^2$.
11. $z = \frac{1}{2}a\log(x^2+y^2)$
$$+ \sqrt{1-a^2}\tan^{-1}\frac{y}{x} + c.$$
12. $x^2 + y^2 + z^2 = z\phi\left(\frac{y}{z}\right)$.
13. $(a-1)z(x+ay+b) - a^2 = 0$.
14. $z = axy + a^2(x+y) + b$.
15. $xz + x\sqrt{x^2+y^2+z^2} = \phi\left(\frac{y}{x}\right)$.
16. $2az = (x+ay)^2 + b$.
17. $(x-u)^3(x+y+z+u)$
$$= \phi\left(\frac{x-y}{z-u},\ \frac{y-z}{z-u}\right).$$
18. The family of planes
$$z = ax + \sqrt{k^2-1-a^2}\,y + b;$$
for b, a fixed constant, the corresponding planes envelop the circular cone
$$(k^2-1)(x^2+y^2) = (z-b)^2,$$
whose vertex is at the point $(0, 0, b)$.

This equation is a particular solution for all values of b.

19. The surfaces of revolution
$$x^2 + y^2 = \phi(z).$$

20. The family of planes
$$z = ax + by + \sqrt{(k^2+c^2)a^2 + k^2b^2 + k^2},$$
where the fixed points are $(c, 0, 0)$ and $(-c, 0, 0)$, and the constant product is k^2. These envelop the ellipsoid of revolution
$$k^2x^2 + (k^2+c^2)(y^2+z^2) = k^2(k^2+c^2).$$

Section 85

4. $x = \phi(z) + \psi(x + y + z)$.
5. $x = \phi(z) + \psi(y)$.
6. $y = x\phi(ax + by + cz) + \psi(ax + by + cz)$.

Section 86

3. $z = \dfrac{xy^3}{3} + y^2\phi(x) + \psi(x)$.
4. $z = \dfrac{x^2y^2}{4} + \phi(x) + \psi(y)$.
5. $z = \dfrac{x^2y}{2} - xy + \phi(y) + e^{-x}\psi(y)$.

Section 88

1. $z = \phi(y + x) + \psi(y + 2x)$.
2. $z = \phi_1(y) + \phi_2(y + 5x) + \phi_3(y + 2x)$.

Section 89

1. $z = \phi(y + x) + x\psi(y + x)$.
2. $z = \phi_1(y + x) + \phi_2(y - x) + x\phi_3(y - x)$.
3. $z = \phi_1(x + y) + x\phi_2(x + y) + \phi_3(x - y) + x\phi_4(x - y)$.

Section 91

4. $z = \phi(y - x) + \psi(2y + 3x) + \frac{1}{2}\sin(x + y) + \frac{1}{8}x^4y + \frac{1}{80}x^5$.
5. $z = \phi(y + 2x) + \psi(y - x) + \frac{1}{12}(2x^3 + y^3)$.

Section 92

3. $z = \phi(y - x) + e^{-2x}\psi(2x + y) - \frac{1}{10}e^{2x+3y} - \frac{1}{6}\cos(y + 2x)$.
4. $z = e^x\phi(y) + e^{-x}\psi(x + y) + \frac{1}{2}\sin(x + 2y) - xe^y$.

Section 93

2. $z = \phi(xy) + \psi\left(\dfrac{y}{x}\right)$.
3. $z = \phi(x^2y) + x\psi(x^2y) + \dfrac{x^3y^4}{30}$.
4. $z = \phi(x^2 + y^2) + \psi(x^2 - y^2)$.

Section 94

1. $z = \dfrac{x^2}{2}\log y + xy + \phi(x) + \psi(y)$.
2. $z = \phi(y + 3x) + \psi(y - 2x) + \frac{1}{6}x^3y + \frac{1}{24}x^4$.
3. $z^2 = x^3y^2 + x\phi(y) + \psi(y)$.
4. $z = \phi(xy) + \psi(x + y)$.
5. $z = \frac{1}{2}x^2y\log x + x^2\phi(y) + \psi(y)$.
6. $z = \phi(x + y) + e^{3x}\psi(y - x) - ye^{x+2y}$.
7. $z = \phi(xy) + x\psi\left(\dfrac{y}{x}\right) + (x - 1)y\log x$.
8. $z = x^{-1}\phi\left(\dfrac{y}{x}\right) + x\psi\left(\dfrac{y}{x}\right)$.
9. $z = x^3y^2 + \phi(y)\log x + \psi(y)$.
10. $z = \phi(x) + \psi(x + y) + (x + y)\log y$.

INDEX

The numbers refer to pages.

The following abbreviations are used: c. c. ≡ constant coefficients. d. e. ≡ differential equation. l. ≡ linear. o. ≡ ordinary. p. ≡ partial. t. ≡ total.

Additive constant, 2.
D'Alembert, 233.
Applications, 31–48, 55, 60, 67, 68, 74, 117–122, 146–148, 162, 163, 214, 229.
Arbitrary constant, 2.
Auxiliary equation, 92, 240, 246.

Bernoulli, 20.
Bernoulli's equation, 20.

Cauchy, 91, 92, 113, 164, 202.
Cauchy's linear equation, 113, 178.
Cayley, 75.
Characteristic equation, 92.
Charpit, 218.
method of Lagrange and, 215.
Clairaut, 25, 56.
Clairaut's equation, 56, 59.
extended, 225.
Commutative operations, 97.
Complementary function, 91, 239.
Complete l. d. e., 90.
Complete solution of o. d. e., 5.
of p. d. e. of first order, 213.
Condition for exactness of o. d. e. of first order, 9.
for exactness of l. o. d. e., 138.
for integrability of t. d. e., 77.
for relation between functions, 253.
for repeated roots of an algebraic equation, 65.
Consecutive points, 70.
Curve of pursuit, 147.
Cuspidal locus, 70.

Darboux, 75, 202.
Degree of d. e., 2.
Derivation of o. d. e., 3.
of p. d. e., 196–201.
Differential equation, 1.
of a family of curves, 31.
of simple harmonic motion, 118.
Discriminant, 65.
relation, 65.

Envelope, 62, 63.
Essential arbitrary constants, 3.
Euler, 25, 91, 192.
factor or multiplier, 25.
Exact differential, 8.
d. e., 8, 11, 137.
Existence theorem for o. d. e., 164.
for p. d. e., 203.

First integral, 52, 230.
Functional determinant, 253.

Gauss, 192.
Gauss's equation, 193.
General integral, 6.
General plan of solution of o. d. e. of first order, 9.
of higher order, 131.
General solution of o. d. e., 5, 167.
of p. d. e., 203.
of p. d. e. of first order, 213.
General summary, 254.
Geometrical significance, 31, 61–63, 66, 69–71, 86, 157, 167, 203, 204, 213.

Homogeneous, function, 14.
l. o. d. e., 89.
l. p. d. e. of first order, 205.
l. p. d. e. with c. c., 239–245.
o. d. e. of first order, 14.
t. d. e., 81.
Hypergeometric series, 194.

Integrable t. d. e., 76–84, 86.
form of solution, 76, 85.
method of solution, 80.
Integral, 6.
curve, 31.
surface, 86.
Integrating factor, 8, 25, 141, 209.
by inspection, 23.
of o. d. e. of first order, indefinite in number, 8.
Integration in series, 164–195.
of o. d. e. of first order, 169.
of o. d. e. of higher order, 177.
Intermediary integral, 231.

Jacobian, 253.

Kowalewski, 202.

Lagrange, 75, 103, 205, 213, 215, 218.
method of, 205.
Lagrange and Charpit, method of, 215.
Legendre, 114.
Legendre's linear equation, 114.
Leibnitz, 25.
Linear o. d. e., 89.
Cauchy's, 113, 178.
complete, 90.
general, 89.
homogeneous, 89.
Legendre's, 114.
of first order, 18.
of second order, 123–130.
reducible to equations with c. c., 113, 114, 126, 127.
simultaneous, with c. c., 150.
with c. c., 91–122.
Linear p. d. e., general, 238.
"homogeneous," with c. c., 239–245.
non-homogeneous, with c. c., 246.
of first order, 200, 205.
Linear p. d. e., of second order, 230.
reducible to equations with c. c., 249.
Linearly independent functions, 90.
Liouville, 142.

Monge, 230.
Monge's equations, 231.
method, 230.

Nodal locus, 69.
Non-homogeneous l. p. d. e. with c. c. 246.
Non-integrable t. d. e., 85, 86.
Non-linear p. d. e. of first order, 211–226.

Order of d. e., 2.
Ordinary d. e., 1.
of first order and first degree, 7–30.
of first order and higher degree, 49–75.
of higher order, 89–148.
reducible to l. d. e. of first order, 20.
system of, 149–163.
Orthogonal trajectories, 38, 158.

Partial d. e., 1.
of first order, 205–229.
of higher order, 230–252.
Particular integral of o. d. e., 6.
of l. o. d. e. in general, 103–105, 125.
of l. o. d. e. with c. c., 97–113.
of l. p. d. e. with c. c., 243, 247.
Particular solution of o. d. e., 5.
of p. d. e. of first order, 213.
Picard, 75, 164.
Point of general position, 32.
Primitive of o. d. e., 4.
of p. d. e., 196, 199.

Quadrature, 6.

Reduction of d. e. to equivalent system, 160.
Regular function, 203.
Riccati, 173.
Riccati's equation, 28, 173.
analogy to l. d. e., 176.
Roots of auxiliary equation repeated, 93, 240.
complex, 94, 242.

Separation of variables, 13, 25.
Series, integration in, 164–195.
Single-valued function, 165.
Singular point of an equation, 203.
Singular solution of o. d. e., 63, 66–75, 168.
  of p. d. e. of first order, 213.
Solution of d. e., 2, 164, 202.
Summary, general, 254.
Symbol D, 89, 238.
  $\mathfrak{D}$, 114, 238.
Symbolic operator $(D-a)$, 96.
System of d. e., 149–163.
  general method of solution, 149.
  of l. o. d. e. with c. c., 150.
System of o. d. e. of first order, 153.
  of t. d. e., 159.

Tac-locus, 71.
Tangent, 70.
Total d. e., 76–88.
  method of solution, 80–85.
  more than three variables involved, 83.
  simultaneous, 159.
Trajectory, 38.

Ultimate points of intersection, 61.
Undetermined coefficients, method of, 107.

Variation of parameters, 103.

REFERENCES

D'Alembert, 233.
Boole, 101, 169, 174, 230.
Cauchy, 113.
Cayley, 75.
Chrystal, 75.
Craig, 91.
Darboux, 75.
Fine, 75.
Forsyth, 174, 230, 244.
Gauss, 195.
Glaisher, 75.
Goursat-Bourlet, 202, 203, 213.
Hermite, 91.
Hill, 72.
Jackson, 122.
Johnson, 174, 244.
Lie, 25, 125.
Liebmann, 75, 168, 169.
Lobatto, 101.
Marie, 233.
Murray, 164, 244.
Picard, 75, 164, 165, 169, 202.
Schlesinger, 125, 141, 164, 183, 185, 194.
Tait and Steele, 45.

Spotted something that isn't quite right?

Let us know and we'll fix it for future reprints. And as a thank you for your diligence, we'll send you a free Sothis book of your choice.

publisher@sothis.press

This page intentionally left blank

Made in the USA
Monee, IL
08 November 2026

68113594R00166